Probability

GRACE E. BATES

Mount Holyoke College

ADDISON-WESLEY · READING, MASSACHUSETTS

Second printing—January 1967

ADDISON-WESLEY PUBLISHING COMPANY, INC.
READING, MASSACHUSETTS · Palo Alto · London
NEW YORK · DALLAS · ATLANTA · BARRINGTON, ILLINOIS

PREFACE

The purpose of the present booklet is to present a short unit in probability theory as a model for experiments resulting in one of a finite number of outcomes. This theory, called a probability theory for *finite* spaces, is remarkably simple in its formulation and in its demands on mathematical background, requiring far less scaffolding, for example, than that needed for Euclidean Geometry. The section on applications gives some indication of how even this restricted theory is used in decision-making problems of industrial sampling, medical studies, etc.

In some of the problems, use is made of elementary combinatorial notions involving the number of arrangements and the number of subsets having a specified number of objects taken from a given set. Appendix 1 contains a brief presentation of such counting techniques for the benefit of the reader unfamiliar with these ideas. There is also a short discussion in Appendix 2 of matrices, in particular of the multiplication of matrices which is employed in Section 8. The material of Appendix 2 is essential background for the development of this section.

The material of Sections 2 through 7 is based on Chapter 12 of the text, *Modern Algebra, Second Course,* by Johnson, Lendsey, Slesnick, and Bates (Addison-Wesley, 1962). Chapter 11 of the same text contains a more detailed presentation of counting principles than that given in Appendix 1 of this booklet.

The introductory treatment of finite Markov chains, which is the subject of Section 8, follows the algebraic approach to this topic given in Kemeny, Snell, and Thompson's *Introduction to Finite Mathematics* (Prentice-Hall, 1957), and the reader of this booklet is instructed to consult this text for a more complete discussion of the theory and applications of finite Markov chains.

On the basis of this elementary excursion into probability theory, it is hoped that interested students may decide at a later time to pursue the subject in more generality, including the field of mathematical statistics. The booklet may be found useful by college students about to embark on a junior year probability course in which material on finite spaces is either presupposed or given only brief coverage.

G. E. B.

South Hadley, Massachusetts
July 1965

CONTENTS

1. INTRODUCTION

Probability theory, like geometry, began as an empirical study consisting of a collection of solutions to everyday problems. The questions to which probability sought answers were of the sort, "How often will this event happen when a like experiment is performed many times?" Gamblers on the continent of Europe and eighteenth-century Masons in England considering life expectancy were two groups of people interested in a systematic analysis of answers to such questions.

In its relatively short history, the theory of probability has proved useful in many areas, notably in the physical and biological sciences, and increasingly in business, economics, sociology, and psychology.

The modern theory of probability is a branch of mathematics, just as geometry, algebra, and number theory are. In its present form it starts with a set of undefined terms and axioms from which theorems are deduced and additional concepts defined. In this book, we shall limit ourselves to that part of the theory having to do with finite probability spaces, which furnish models for experiments resulting in one of a finite number of outcomes. The more general theory, while requiring more careful formulation and deeper mathematical concepts, may essentially be grounded in the special theory of finite spaces.

2. FINITE SAMPLE SPACES AND PROBABILITY MEASURES

We start with the notion of an experiment being performed with a set of objects, in which certain specified outcomes are of interest. Examples of such experiments are;

EXAMPLE 1. A die is tossed and the number of dots showing on the top face counted.

EXAMPLE 2. One of several balls painted red or white is drawn from an urn and its color noted.

EXAMPLE 3. An individual is selected from a group of people and his opinion asked concerning a certain proposal.

Each experiment gives rise to a set of possible outcomes. The set of all possible outcomes of an experiment is called a *sample space* or *probability space* of the experiment, and each possible outcome is called a *point* of the sample space. In Example 1 above, a sample space is the set {1, 2, 3, 4, 5, 6} ; for Example 2, a sample space {red ball, white ball} is appropriate; and in Example 3, sets S_1 = {for, against} or S_2 = {for, against, undecided} suggest themselves.

It is important to note that in the mathematical theory under consideration, the terms *sample space* and *sample point* are undefined and form the starting point of the theory. In applying the theory then, the first task is to interpret in the light of the actual experiment what is to play the role of sample space and what are the corresponding sample points.

To each possible outcome of an experiment we assign a number or *weight*, which is to be used to make probability statements about the experiment. In Example 1 above we ordinarily think of assigning equal weight to each of the six possible outcomes. That is, we consider each of the six numbers as having its share, one-sixth, of the total weight and use $w(i) = \frac{1}{6}$ for $i = 1, 2, 3, 4, 5, 6$ to denote this apportionment. With this allotment we may agree that the probability of an odd number (i.e., 1, 3, or 5) on the top face will be $3(\frac{1}{6})$ or $\frac{1}{2}$, which will also be the probability of an even number (2, 4, or 6). On the other hand, should there be reason to doubt the balance of the die or the method of throwing it, it is conceivable that an assignment of weights such as $w(1) = w(6) = \frac{1}{4}$, $w(2) = w(3) = w(4) = w(5) = \frac{1}{8}$ might be suggested. In other words, in the formulation of this part of the theory we merely require the following:

> **Fundamental postulate of weights.** To each point o_i of a sample space, we assign a nonnegative number $w(o_i)$, called the *weight* of o_i, in such a way that the sum of the weights of all points of the sample space is 1.

With this stipulation in mind, if the composition in the urn in Example 2 above is, say, 20 red balls and 40 white ones and if the collection of balls is considered to be "thoroughly mixed," we would probably assign to the outcome "red ball" the weight $\frac{20}{60}$ or $\frac{1}{3}$ and to the outcome "white ball" the weight $\frac{2}{3}$. Example 3 poses more of a problem in abstracting from the actual situation a sensible assignment of weights, but we might have recourse to the point of view of an impartial observer and suggest for space S_1 a weight of $\frac{1}{2}$ to each of the outcomes "for" and "against" or, alternatively, for S_2 use weights of 0.45 for "for," 0.45 for "against," and 0.10 for "undecided." There are certainly other possible assignments.

The assignment of weights to the possible outcomes, once the latter are identified in an actual experimental situation, is again extramathematical, a matter of convenience and purpose in conducting the particular experiment. The usefulness of probability theory and its applications, however, lies in part in a comparison between hypothetical or assumed weights in a probability situation with the actual relative frequencies or proportion of observed occurrences of a particular event. Thus in the case of a suspect die, an actual experiment of repeated throws may furnish apparent evidence or lack of evidence to confirm this suspicion when the observed numbers of occurrences of dots on its top face are compared with those predicted by theoretical probability considerations of a fair or balanced die.

To summarize, we have in our formulation of a theory of probability the (undefined) notion of *finite sample space* and its *points*, and an assignment of nonnegative numbers or *weights* to each point of the space in such a way that the sum of these weights for all points of the space is 1. We shall next define the concept of "event." This terminology stems, of course, from the applications of the theory.

Definition of Event. An *event* in a sample space is a subset of the sample space.

In the suggested sample space of Example 1, namely $S = \{1, 2, 3, 4, 5, 6\}$, the following are events: the set A of all top faces which are odd numbers, that is, $A = \{1, 3, 5\}$; the set B of top faces whose numbers are less than 5, or $B = \{1, 2, 3, 4\}$. We shall also agree to consider the space S itself as an event; for example, in Example 1, S is the event consisting of the numbers 1 through 6 inclusive. Also, in keeping with the convenient notion of *empty set* or *null set*, we allow as an event C: "a number less than 2 and greater than 5." Of course, $C = \emptyset$, the empty subset of S. An illustration of an event from Example 3, using sample space S_2: {for, against, undecided}, might be the event D: "not opposed," consisting of $D =$ {for, undecided}.

In any finite sample space we have as rather special events those subsets each of which contains just one point of the sample space. Thus, in the die-throwing example, the outcome consisting of a top face number "3" is itself an event which might be denoted by $\{3\}$. In the language of sets, these are called *unit sets*, sets consisting of single elements.

We shall complete the mathematical formulation of this theory of finite probability spaces with the following definition.

Probability of an Event. Given a finite sample space S of points $o_1, o_2, \ldots, o_n$ to which weights $w(o_i)$ have been assigned with $w(o_i) \geq 0$ and $\sum_{i=1}^{n} w(o_i) = 1$, the *probability of an event* E of S is given by

$$P(E) = \sum_{o_i \text{ in } E} w(o_i).$$

That is, *the probability of an event E of S is the sum of the weights attached to the points of E. In particular, we require that the probability of the event $\emptyset$, which is the empty subset of S, be* 0.

Some additional problems illustrating the ideas in this section are given below.

PROBLEM 1. Two pennies and a nickel are tossed. (a) What is the probability that all three will show heads? (b) What is the probability that both pennies will show the same side; that is, both heads or both tails?

Assuming that we cannot distinguish between the two pennies, there are six possible observable outcomes of this experiment:

$$S = \{HHH, HHT, HTH \text{ (or } THH), HTT \text{ (or } THT), TTH, TTT\}.$$

In the listing of points of this sample space S, the first two letters refer to the way the two pennies fall and the third to the way the nickel falls. We have placed THH in parentheses after HTH to remind us that HTH and THH refer to the same observable outcome (head on one penny, tail on the other, and head on the nickel). Similarly, HTT and THT are indistinguishable outcomes.

In assigning weights to the six points, an assignment of $\frac{1}{6}$ to each point is certainly compatible with and satisfies the sole requirement of the theory that the total of the weights be 1. However, in an actual experiment, it is not difficult to convince oneself that the outcome designated by HTH (or THH) has a better chance of occurring than, say TTH, or any other outcome in which the two pennies both show the same face. To see this, imagine one of the pennies painted red. Thus the assignment suggested by the experiment is

$$w(HHH) = w(HHT) = w(TTH) = w(TTT) = \tfrac{1}{8},$$
$$w(HTH \text{ or } THH) = w(HTT \text{ or } THT) = \tfrac{2}{8} \quad \text{or} \quad \tfrac{1}{4}.$$

Note that the total of these weights is again properly 1. The probabilities required in questions (a) and (b) are then:

(a) $P\{HHH\} = \frac{1}{8}$,

(b) $P\{HHH, HHT, TTH, TTT\} = \frac{4}{8}$ or $\frac{1}{2}$.

Of course, the above assignment of weights might still be unsatisfactory in furnishing an adequate model if one or more of the coins were decidedly unbalanced or, perhaps, two-headed!

A situation similar to that of Problem 1 arises when we conduct experiments involving a throw of two or more dice—we mentally paint them different colors and then analyze the problem of balanced dice by assuming that each face of each die is equally likely to show.

It is often possible to classify outcomes of an experiment in different ways. It is for this reason that one properly speaks of *a* sample space of an experiment rather than *the* sample space. The following problem illustrates this possibility of more than one analysis of possible outcomes. In this discussion we shall make use of certain counting procedures which are briefly described in Appendix 1.

PROBLEM 2. Two cards are drawn in succession from a well-shuffled deck; the first card is not replaced before the second is drawn. What is the probability that one card is a heart and the other a club?

This experiment involves a sequence of two actions, and hence a natural choice of sample space is the set S of all ordered pairs of the deck, the first element of the ordered pair representing the first card drawn and the second element, the second card drawn. For example, (H_2, D_K) and (D_K, H_2) are different points of S, where H_2 is the two of hearts and D_K is the king of diamonds. The number of points of S is $52 \cdot 51$ or 2652 [cf. the Sequential Counting Principle (SCP) in Appendix 1]. Assuming equal weights, $\frac{1}{2652}$ to each possible point of S, we count the number of points (ordered pairs) in the event E in question. There are $13 \cdot 13 = 169$ ordered pairs of type (H, C) and also 169 ordered pairs of type (C, H), so that there are 338 points of S in the event E consisting of one heart and one club. Hence $P(E) = \frac{338}{2652}$ or $\frac{13}{102}$.

As an example of a second choice of probability space, say S', for this experiment, we may consider the points of S' as *subsets* of two cards each (i.e., unordered pairs), since the actual performance of the experiment suggested is one way of constructing a list of all possible subsets of two cards each from the deck. The number of points of S' is $\binom{52}{2} = 1326$, using this binomial coefficient (cf. Appendix 1) to count the number of 2-subsets from the set of 52 cards. Assigning weight $\frac{1}{1326}$ to each point of S', we count the number of points in the event $E' = \{$subsets consisting of one heart, one club$\}$. There are $\binom{13}{1}\binom{13}{1} = 169$ points in E' and we have $P(E') = \frac{169}{1326} = \frac{13}{102}$, as before.

The first analysis of outcomes, which results in space S, is said to be a *finer* analysis than that leading to the space S'. Note that the finer analysis would be needed to answer such questions as "What is the probability of drawing first a heart, then a club?" The reader may verify that the answer to this question, using space S, is $\frac{169}{2652}$ or $\frac{13}{204}$.

EXERCISES

In Exercises 1 through 9, first specify a sample space S underlying the experiment, then assign weights to its points on the basis of equally likely outcomes. After this has been done, describe the event in question as a subset of the sample space and compute its probability. The first five exercises contain suggested analyses for specifying a sample space. First, a quick review of Appendix 1 is recommended.

1. A penny is tossed three times. Find the probability of

(a) 3 heads
(b) 2 heads and 1 tail
(c) 1 head (i.e., exactly 1)
(d) no head
(e) at least one head
(f) at most one head

[List the points of S as ordered triples; there are 8 of these.]

2. Two cards are drawn from a well-shuffled deck of 52 cards.

(a) What is the probability that both are jacks?
(b) What is the probability that both are queens?
(c) What is the probability that one is a jack and one a queen?

[Analyze the space into possible 2-subsets of cards from the deck.]

3. A committee of 4 is to be chosen at random from a group of 10 students. What is the probability that a particular student A in the group will be elected to serve on the committee? [Analyze the space into possible 4-subsets from the 10 people.] It is customary to use the phrase "at random" to imply an assignment of equal weights to each of the possible outcomes.

4. An executive committee of three, consisting of a president, vice-president, and secretary-treasurer is to be chosen at random from a panel of 5 students, one of whom is Susie Smith.

(a) What is the probability that Susie will become president?
(b) What is the probability that Susie will be put on the committee?

[Analyze the space into all possible permutations of 3 objects from the set of 5 objects.]

5. Three boys and 4 girls are lined up in a row for a picture. What is the probability that the girls and boys alternate? [Analyze the space into all possible permutations of 7 objects out of the group of 7 people.]

6. A lot of 50 items contains 5 defective items and 45 nondefective ones. A random sample of 3 is selected from the lot and each item of the sample tested. What is the probability that all 3 items in the sample are defective? none are defective? at least one is defective?

In this exercise, we may consider a sample space S_1 as the set of all possible 3-subsets drawn from the set of 50 items *or* (cf. Problem 2) a sample space S_2 of ordered triplets of items, without repetition, drawn sequentially from the set of 50 items. Such a sampling procedure is termed sampling *without replacement*, since the successive items of the sample are not replaced before the next one has been drawn.

7. An urn contains 100 balls, of which 20 are white, 20 are green, 30 blue, and 30 red. Two balls are drawn at random from the urn without replacement. What is the probability that both are green? one is green and one is blue? neither is white?

8. Using only numerals from the set $\{1, 2, 3, 4\}$ and not allowing repetitions, a list of all possible three-digit numbers is formed and each number is written

on a cardboard disc. From the set of discs, one disc is drawn at random. What is the probability that the number on the disc selected

(a) begins with the digit 1?
(b) begins and ends with an even digit?
(c) is an odd integer?

9. Answer the three parts of Exercise 8 if the list of three-digit numbers formed from the set $\{1, 2, 3, 4\}$ also allows repetitions of digits, that is, numbers such as 233 and 444.

10. Suppose that a die has its faces weighted so that the probabilities are in the same ratio as the numbers on the faces. (That is, a 6 is six times as likely to show as a 1 and twice as likely as a 3, etc.) When this die is tossed once, what is the probability that an odd number shows? that an even number shows?

3. COMBINATION OF EVENTS

Since events are defined to be subsets of a given set S, the sample space, the algebra of the combination of sets is relevant to a study of the combinations of events in probability theory. A useful device for picturing the relationship between various subsets of a set is the so-called Venn diagram, named after an English logician, John Venn (1834–1883). As an illustration, let us consider a sample space of ten points denoted by $S = \{a, b, c, d, e, f, g, h, i, j\}$. We will assume that to each of these points there is assigned a weight, so that we have numbers $w(a), w(b), \ldots, w(j)$ such that each of these is nonnegative and their total is 1.

In Fig. 1, we use a rectangle S to enclose these ten points, and within the rectangle we have drawn a circle A enclosing the four points of the subset (event) A given by $A = \{a, b, c, d\}$. The points of S *not* in A also form a subset of A which we denote by $\overline{A}$ and call *the complement of* A in S. Thus,

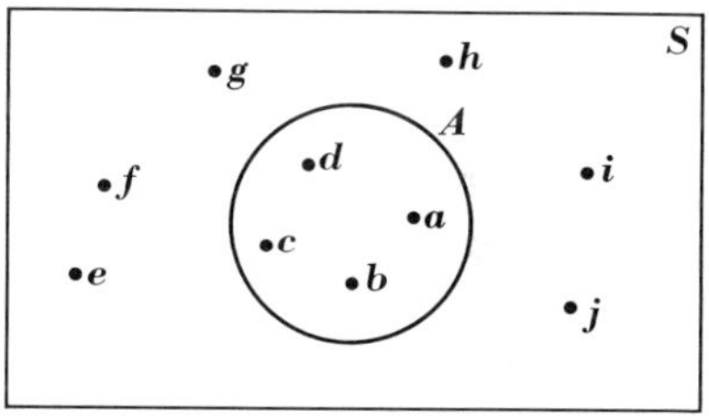

$$\overline{A} = \{e, f, g, h, i, j\}.$$

FIGURE 1

Note that since the total of the weights of all points of S is 1, we have

$$\begin{aligned} P(\overline{A}) &= w(e) + w(f) + w(g) + w(h) + w(i) + w(j) \\ &= 1 - [w(a) + w(b) + w(c) + w(d)] = 1 - P(A). \end{aligned}$$

Since the above argument in no way depends on our particular space S and the number of its points, we have proved the following

Theorem 1. If A is an event of a sample space S, then $P(\overline{A}) = 1 - P(A)$.

The first theorem of our theory, simple as it is, has useful applications. For example, to compute the probability of the event A: "at least one head in three tosses of a coin," it is simpler to compute the probability of the complementary event $\overline{A}$: "no heads in three tosses"; that is, $P(\overline{A}) = \frac{1}{8}$, since only one, namely TTT, of the eight possible outcomes in the three tosses results in no heads. Then $P(A) = 1 - P(\overline{A}) = \frac{7}{8}$.

Venn diagrams are useful also in picturing two or more subsets of a set S. In Fig. 2, for example, we have drawn the Venn diagram for the set S used in Fig. 1 and the subsets $A = \{a, b, c, d\}$ and $B = \{c, d, f, g\}$. The two points in the "overlap" of circles A and B represent the elements c and d of S which are in both A and B; we call the set $\{c, d\}$ the *intersection* of sets A and B and denote this intersection set by $A \cap B$.

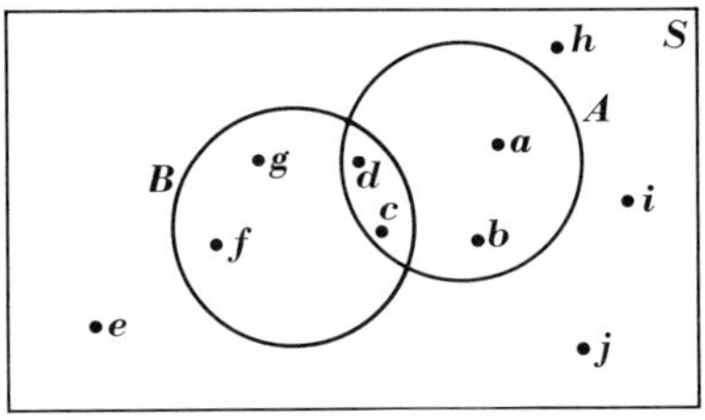

FIGURE 2

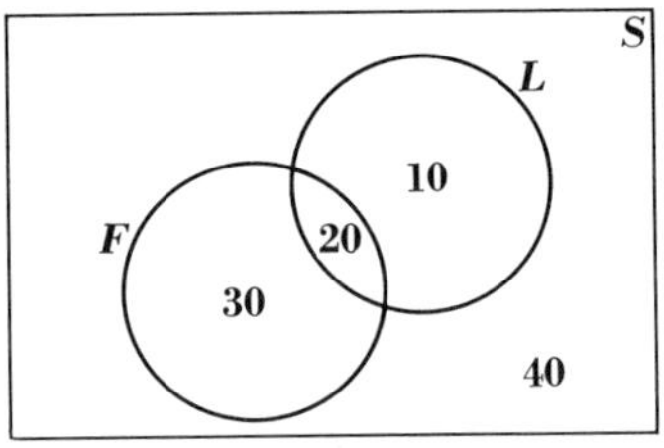

FIGURE 3

Another subset of S related to sets A and B is the set of all elements of S which are in A or in B, including, of course, any elements in both A and B. This set is called the *union* of sets A and B and is denoted by $A \cup B$; thus

$$A \cup B = \{a, b, c, d, f, g\}$$

for the sets A and B of Fig. 2.

How is the probability of the set $A \cup B$ related to that of sets A and B? Because of the overlap of points in both A and B, that is, those in $A \cap B$, the answer is not immediately apparent. Suppose we consider the two numbers $P(A)$ and $P(B)$ which give the probability of set A and set B, respectively. Since each of these numbers is the result of adding the weights attached to points of each subset, what happens when we rather naturally form their sum $P(A) + P(B)$ as a candidate for the answer to the question, "What is $P(A \cup B)$?" On careful examination of what numbers are being combined, we see that the sum $P(A) + P(B)$ counts the contribution of weights of points in $A \cap B$, the intersection set, twice. Thus, we have the general theorem:

Theorem 2 *Addition theorem.* If A and B are events in a sample space S, then $P(A \cup B) = P(A) + P(B) - P(A \cap B)$.

The reader may verify this statement for the events A, B in the sample space S of Fig. 2 by assigning arbitrary weights totaling 1 to the ten points of S, then computing $P(A \cup B)$ directly and by the formula of Theorem 2.

A special case of the Addition Theorem is embodied in the following corollary.

Corollary to Theorem 2. If A and B are events in a sample space S with $A \cap B = \emptyset$, then $P(A \cup B) = P(A) + P(B)$.

The proof of this corollary is left to the reader.

Definition. Events A and B in a sample space S for which $A \cap B = \emptyset$ are called *mutually exclusive events.*

For example, in the experiment of three tosses of a balanced coin, the events A: "exactly one head" and B: "no tails" are mutually exclusive events since the event $A \cap B$ of exactly one head *and* no tails has no points or is what one frequently calls "an impossible event."

The following problems illustrate the use of Venn diagrams in applying the theorems of this section.

PROBLEM 1. The junior class has 100 students of whom 50 take French, 30 take Latin, and 20 take both French and Latin. If a student is chosen at random from the class, what is the probability that the student takes at least one of these two languages? What is the probability that he takes French but not Latin?

In Fig. 3 we have indicated the sample space S of 100 possible student selections by a rectangle and have drawn circles F and L, indicating enclosures for those subsets of students taking French and Latin, respectively. Instead of denoting each student by a point, we have indicated the *numbers* of students in each enclosure. These numbers were entered in the following order: first, we entered 20 for $n(F \cap L)$, the number of students taking both French and Latin. This left 30 students for the rest of F and 10 for the rest of L, making a total of $20 + 30 + 10 = 60$ students taking at least one of the two languages (that is, in $F \cup L$), and hence $100 - 60 = 40$ taking neither language. Under the usual assumption of "random" selection of a student from the group of 100 students, we agree to assign a weight of $\frac{1}{100}$ to each student of the class. The first question requires the computation of $P(F \cup L)$ which we compute directly from Fig. 3 as $(30 + 20 + 10)/100$ or $\frac{3}{5}$. We may also check that $P(F \cup L) = P(F) + P(L) - P(F \cap L) = \frac{50}{100} + \frac{30}{100} - \frac{20}{100} = \frac{3}{5}$. Finally, the second question asks for $P(F \cap \bar{L})$, that is, for the probability of that subset of F not also contained in L. From Fig. 3 we have $P(F \cap \bar{L}) = \frac{30}{100}$ or $\frac{3}{10}$.

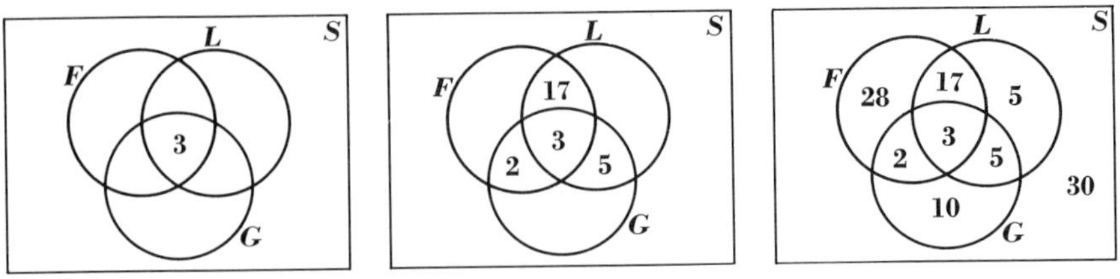

FIGURE 4

PROBLEM 2. Suppose that the following additional information is given about the junior class of Problem 1: 20 students take German, 5 take both French and German, 8 take both German and Latin, 3 take all three languages. What is the probability that a student selected at random (a) takes French but no other of these languages? (b) does not take any one of the three languages? In Fig. 4 we give the construction, in three stages, of the Venn diagram related to this problem. The reader may then verify the following solutions to the two parts of the problem:

(a) $P(F \cap \overline{L} \cap \overline{G}) = \frac{28}{100}$ or $\frac{7}{25}$,

(b) $P(\overline{F} \cap \overline{L} \cap \overline{G}) = P(\overline{F \cup L \cup G}) = \frac{30}{100}$ or $\frac{3}{10}$.

EXERCISES

1. $S = \{o_1, o_2, o_3, o_4, o_5\}$ is a sample space of five points with weights assigned to satisfy the following conditions:

$$w(o_1) = w(o_2) = w(o_3),$$
$$w(o_4) = w(o_5) = 2w(o_1).$$

Two subsets A and B of S are denoted by

$$A = \{o_1, o_3, o_4\},$$
$$B = \{o_2, o_3\}.$$

(a) Find the weights assigned to the points of S.
(b) Make a Venn diagram labeling the points of S and draw circles enclosing the points of the subsets A and B.
(c) Find $P(A \cap B)$, $P(A \cup B)$, $P(\overline{A})$.

2. A die is tossed twice.

(a) What is the probability of a 6 on the first toss?
(b) What is the probability of a 6 on the second toss?
(c) What is the probability of a 6 on at least one of the tosses?

3. From the set of 4 aces, 2 are drawn at random. What is the probability that in the pair selected at least one of the cards is a red ace?

4. A certain class of 80 students boasts 20 honors students and 30 school athletes. Forty students in the class do not indulge in sports and are not honors students. If a student is selected at random to represent the class, what is the probability that he is both an honors student and a school athlete? that he is in at least one of these two special categories?

5. In a group of 100 girls, 40 are blondes and the rest brunettes. Fifty of the girls have blue eyes, the others brown eyes. There are 10 blue-eyed blondes. If a boy dates one of these 100 girls, choosing her name by lot, what are his chances of going out with a brown-eyed brunette?

6. An integer is chosen at random from the first 20 positive integers. What is the probability that the integer chosen is divisible by 4 or by 6? (Try to do this exercise in two ways.)

4. CONDITIONAL PROBABILITY AND INDEPENDENT EVENTS

Consider again Problem 1 of the last section. The Venn diagram of Fig. 5 indicated the number of students taking French, Latin, both French and Latin, and neither of these two languages. We ask the following two questions.

(a) The name of a student is chosen at random from the class list. What is the probability that this student is taking French?

(b) The name of a student is chosen at random from the class list. Upon checking the files, it is found that this student is taking Latin. What is the probability that he is (also) taking French?

The solution to (a) is easily seen to be

$$P(F) = \tfrac{50}{100} \quad \text{or} \quad \tfrac{1}{2}\,.$$

Although the same experiment is performed in (b), additional information is now available about the student selected. The question in (b) might be rephrased as follows: knowing that the student chosen is taking Latin, what is the probability he is taking French? When stated in this way, it is clear that the solution to (b) is

$$\frac{n(F \cap L)}{n(L)} = \tfrac{20}{30} \quad \text{or} \quad \tfrac{2}{3}\,,$$

since of the 30 equally likely choices of a Latin student, 20 are also enrolled in French. Note that the knowledge that the student is taking Latin has made it more probable that he is taking French, the probability having increased from $\frac{1}{2}$ to $\frac{2}{3}$.

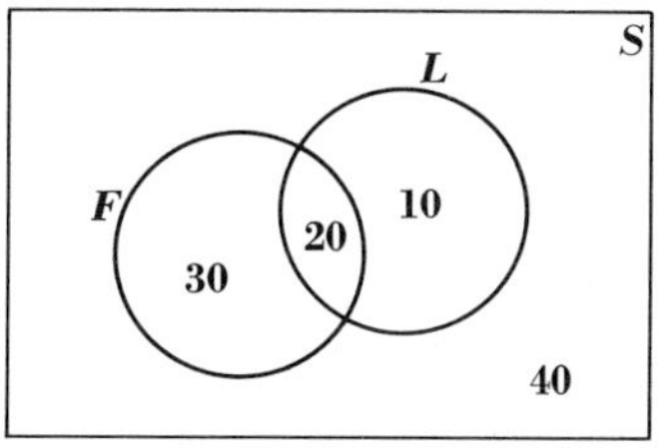

FIGURE 5

This example illustrates an important probability concept, that of conditional probability. The probability $\frac{2}{3}$ in answer to question (b) is the *conditional probability that a student takes French, given that he takes Latin.* The probability $\frac{1}{2}$ obtained as the solution to part (a) is sometimes called the *absolute probability* that the student takes French, since in part (a) no condition was imposed on the random selection of a student from the class. The definition of conditional probability is given below:

Definition of Conditional Probability. If A and B are events in a sample space S, with $P(B) \neq 0$, then *the conditional probability of event A, given event B,* is denoted by $P(A|B)$ and defined by

$$P(A \mid B) = \frac{P(A \cap B)}{P(B)}.$$

That this definition does indeed provide a proper probability assignment restricting ourselves to just the subset B of the sample space S may be seen as follows. To each point o_i of S which lies in the subset B we now give weight $w'(o_i) = w(o_i)/P(B)$, where $w(o_i)$ is the weight originally assigned to o_i in S. It is clear that the new weights $w'(o_i)$ for points of B are thus proportional to the original weights of these points [with constant of proportionality $1/P(B)$] and that the sum of the $w'(o_i)$ for all points o_i in B is just $[1/P(B)] \cdot P(B)$ or 1.

Using the notation of the definition above, we find that the answer to question (b) is

$$P(F \mid L) = \frac{P(F \cap L)}{P(L)} = \frac{\frac{20}{100}}{\frac{30}{100}} = \frac{2}{3}.$$

PROBLEM 1. A student council consists of 15 students of whom 10 are seniors and 5 are juniors. Of the 10 seniors, 5 are girls and 5 are boys; of the 5 juniors, 2 are girls and 3 are boys. One student is chosen by lot from the council. (a) What is the probability that this student is a boy? (b) Knowing that the student chosen is a senior, what is the probability that this student is a boy?

The sample space consists of the 15 possible choices, which we consider to be equally likely. Let B be the event "boy student" and A the event "senior student." From the numbers given we have $n(B) = 8$, $n(A) = 10$ and $n(B \cap A) = 5$. Hence the solutions to the two parts of the problem are:

(a) $P(B) = n(B)/15 = \frac{8}{15}$,

(b) $P(B|A) = P(B \cap A)/P(A) = \frac{5}{15}/\frac{10}{15} = \frac{1}{2}$.

We note in this example that the conditional probability of event B given A is less than the absolute probability of B; that is, the probability of the choice being a boy decreases with the additional information that a senior has been chosen. In our first illustration of this section, added information increased the absolute probability.

An interesting case of conditional probability arises when additional information has no effect on the probability of the event considered. If events A and B of a sample space S are such that $P(A|B) = P(A)$, the event A is said to be *independent* of event B. The following problem illustrates this concept.

PROBLEM 2. Last year's student council consisted of 4 girls and 6 boys among the 10 seniors and 2 girls and 3 boys among the 5 juniors. Answer questions (a) and (b) of Problem 1 for this situation.

Using the notation of Problem 1, we find that for last year's council:

(a) $P(B) = n(B)/15 = \frac{9}{15}$ or $\frac{3}{5}$,

(b) $P(B|A) = P(B \cap A)/P(A) = \frac{6}{15}/\frac{10}{15} = \frac{3}{5}$.

In view of our definition above, we may say that the event "boy student" is independent of the event "senior student."

The complementary event to A, denoted by $\overline{A}$, is "junior student" for the space of this problem. Note that B is also independent of $\overline{A}$, since

$$P(B \mid \overline{A}) = \frac{P(B \cap \overline{A})}{P(\overline{A})} = \frac{\frac{3}{15}}{\frac{5}{15}} = \frac{3}{5},$$

which is the same as $P(B)$.

The fact that both $P(B|A)$ and $P(B|\overline{A})$ are equal to $P(B)$ is what one has in mind when one thinks intuitively about independent events; that is, B has the same probability *whether or not* A occurs. In Problem 2 this is easily seen from the fact that the proportion of boys on the council is the same overall as in the group of senior members and as in the group of junior members.

If event A is independent of B so that $P(A) = P(A|B)$, then

$$P(A) = P(A \mid B) = \frac{P(A \cap B)}{P(B)}$$

or

$$P(A) \cdot P(B) = P(A \cap B).$$

From the last equation, if $P(A) \neq 0$, we may obtain

$$P(B) = \frac{P(A \cap B)}{P(A)} = P(B \mid A).$$

That is, we have shown that if A is independent of B, where neither A nor B are null events (i.e., events with probability 0), then $P(A \cap B) = P(A) \cdot P(B)$ and B is independent of A.

Similarly, starting with $P(A \cap B) = P(A) \cdot P(B)$, with neither $P(A)$ nor $P(B)$ equal to zero, we may show that this equation implies that A is independent of B and B independent of A, in which case we say that A and B are *independent events*. These remarks constitute a proof of the following theorem.

> **Theorem 3.** If A and B are events having nonzero probabilities in a sample space S, then A and B are *independent events* if and only if $P(A \cap B) = P(A) \cdot P(B)$. In words, two non-null events are independent if and only if the probability of their joint occurrence equals the product of their separate probabilities.

Note. It is customary to call events A and B independent if either has zero probability, because then $P(A \cap B) = 0$ and the equation $P(A \cap B) = P(A) \cdot P(B)$ reduces to the true statement $0 = 0$.

PROBLEM 3. In the experiment of tossing three balanced coins, let A be the event "at least two heads" and B the event "same face on all three coins." Are events A and B independent?

Considering as a sample space for this experiment the set of eight ordered triples of letters H and T (standing for heads and tails, respectively), the subsets A, B, $A \cap B$, are:

$$A = \{HHH, HHT, HTH, THH\}, \qquad B = \{HHH, TTT\},$$
$$A \cap B = \{HHH\}.$$

Hence $P(A) = \frac{4}{8}$ or $\frac{1}{2}$; $P(B) = \frac{2}{8}$ or $\frac{1}{4}$; $P(A \cap B) = \frac{1}{8}$. Since $P(A \cap B) = P(A) \cdot P(B)$, we conclude from Theorem 3 that A and B are independent events.

The reader may check that in this same experiment, the events C: "at most two heads" and B: "same face on all three coins" are not independent events, since $P(B \cap C) \neq P(B) \cdot P(C)$.

If two events are not independent, then they are said to be *dependent events*. In view of Theorem 3, two events are *dependent* if and only if the probability of their joint occurrence does *not* equal the product of their separate probabilities. The independence of events A and B but dependence of events B and C for the experiment of Problem 3 illustrate the fact that the idea of independence in the probability sense (sometimes called statistical independence) is not necessarily in agreement with our intuitive notion of independence.

The concept of independent events may be extended to cover three or more events. Thus we say that a set of three events is a *completely independent set* if all pairs of the events are independent and if also the probability of the joint occurrence, that is, of the intersection of all three events, is the product of the separate probabilities of these events. A set of four events is said to be completely independent if all pairs and all triples of the events are independent, and if also the probability of the joint occurrence of all four events is the product of the separate probabilities of the four events, and so on for five or more events.

A set of events is said to be *pairwise independent* if each pair of these events is independent. It is a curious fact that a set of events may be pairwise independent without being completely independent. This case is illustrated in Exercise 5 below.

Beginning students of probability theory sometimes confuse the two notions of *mutually exclusive* events and *independent* events. We leave it to the reader to verify the following statement: If two non-null events are independent, then they are *not* mutually exclusive, and two non-null events which are mutually exclusive cannot be independent.

EXERCISES

1. Prove the statement concerning mutually exclusive and independent events made at the end of this section.

2. In a class of 80 students of which 20 are honors students, 30 are athletes, and 40 are neither athletes nor honors students, a student is chosen at random.

(a) What is the probability that the student chosen is an athlete, given that he is an honors student?

(b) What is the probability that the student selected is an honors student, given that he is an athlete?

(c) Are the events "honor student" and "athlete" independent in this experiment?

3. A green die and a red die are each tossed once.

(a) Find the conditional probability of obtaining a sum greater than 9, given that the red die shows a 6.

(b) Find the conditional probability that the sum is less than 6, given that the green die shows a 2.

(c) Find the conditional probability of obtaining the sum 11, given that the red die shows a number less than 5.

(d) In each of parts (a) through (c), how does the given additional information affect the probability of the event in question?

4. (a) Two fair coins are tossed. Let A be the event "not more than one head" and B the event "at least one of each face." Show that A and B are dependent events.

(b) Three fair coins are tossed. Let A' and B' be the events of this space corresponding to the verbal descriptions of A and B in part (a). Are A' and B' independent events?

5. A number is chosen at random from the set of four code numbers, $\{100, 010, 001, 111\}$. Let E be the event "first digit 1," F the event "second digit 1," and G the event "third digit 1." Show that the sets of events E, F, G is a pairwise independent set but not a completely independent set.

6. Consider the set of eight cards consisting of just the four jacks and the four queens of a bridge deck and suppose that two cards are drawn at random from this set of jacks and queens.

(a) Define a sample space S of the experiment and determine $N = n(S)$. In answering the following questions, you may find it worthwhile to list all these points of S.

(b) Find the probability that the cards are both jacks, given that (at least) one is a jack.

(c) Find the probability that both cards are jacks, given that one is a red jack.

(d) Find the probability that both are jacks, given that one is the jack of hearts.

7. The table below shows a two-way classification of a group of 100 people according to sex and political affiliation.

	Democrats	Republicans	Total
Men	39	21	60
Women	26	14	40
Total	65	35	100

Suppose that a person is chosen at random from this group, and let D stand for the event "Democrat" and M the event "male." Are the events D and M independent? How about events D and $\overline{M}$? $\overline{D}$ and M? $\overline{D}$ and $\overline{M}$?

8. According to the American Experience Mortality Table, of 100,000 persons living at age ten, 95,550 live to be sixteen years of age and 69,804 live to be fifty years of age. If you are sixteen years of age, what are your chances of living to be fifty years old?

5. TREE MEASURE

PROBLEM 1. Two urns are placed on a table (Fig. 6). While the urns look alike, urn 1 contains 3 red balls and 6 white ones, whereas urn 2 contains 4 red balls and 4 white balls. A person chooses one of the urns at random and draws a ball. What is the probability that he draws a red ball?

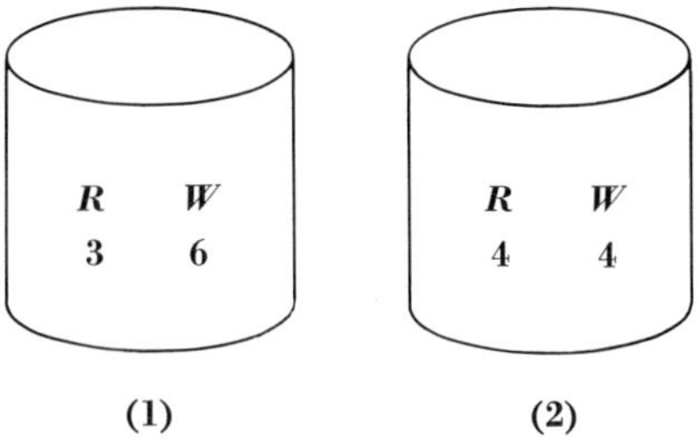

FIGURE 6

This experiment involves a sequence of two actions: choose an urn, then draw a ball from this urn. Since we are interested only in the color of the ball drawn, a sample space relevant to the problem is

$$S = \{1R, 1W, 2R, 2W\},$$

where $1R$ denotes the outcome

"choose urn 1 and draw a red ball from it," etc. It is clear that an assignment of equal weights to these four points will not be appropriate, because of the different composition of colored balls in the two urns. At the same time, weighting each of the possible draws of a ball as $\frac{1}{17}$ would suggest a different experiment in which all the balls are placed in one urn. What we really require is a weight assignment such as that indicated in the scheme below:

Points	Weights	
$1R$	x	total $\frac{1}{2}$
$1W$	$2x$	
$2R$	y	total $\frac{1}{2}$
$2W$	y	
		total 1

That is, we weight $1R$ and $1W$ in the ratio $1:2$, since this is the ratio of red to white balls in urn 1; similarly, the weights for $2R$ and $2W$ should be the same, since there are four balls of each color in urn 2. We want the total of weights in each of these two groupings to be $\frac{1}{2}$, since in each case this total gives the probability of selecting a particular urn at the start. By simple algebra we find that $x = \frac{1}{6}$, $2x = \frac{1}{3}$, $y = \frac{1}{4}$, and hence the answer to the probability question asked, the probability of drawing a red ball, is

$$P\{1R, 2R\} = x + y = \tfrac{1}{6} + \tfrac{1}{4} = \tfrac{5}{12}.$$

Problem 1 can be more easily attacked by using a *tree diagram* as shown in Fig. 7.

In Fig. 7, we have drawn two arrows from "start," one to the outcome representing the choice of urn 1 and the other to the outcome representing the choice of urn 2. We have labeled each of these arrows with the number

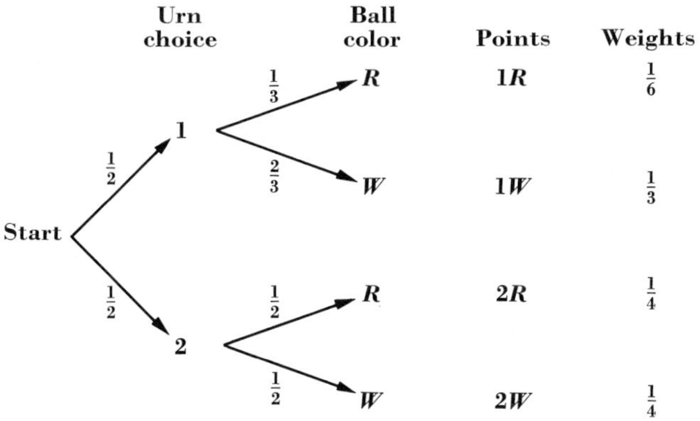

FIGURE 7

$\frac{1}{2}$, representing the "random" choice of urn in the first stage of our experiment. The two arrows from urn choice 1 to R and W represent the two possible outcomes of a draw from this urn and, similarly, the two arrows from urn choice 2. The numbers $\frac{1}{3}$ and $\frac{2}{3}$ for arrows from 1 to R and W, respectively, take into account the fact that urn 1 has $\frac{1}{3}$ of its contents red balls and $\frac{2}{3}$ white. The numbers $\frac{1}{2}$ on each of the arrows from 2 to R and W are similarly explained. The points of the underlying sample space S of this sequence of two actions are denoted by labeling the paths from start through the two stages of the experiment. The weights written in the last column were obtained by multiplying the numbers assigned to the two branches of the corresponding path; thus the weight $\frac{1}{6}$ for point $1R$ is $\frac{1}{2} \cdot \frac{1}{3}$; the weight for $1W$ is, similarly, $\frac{1}{2} \cdot \frac{2}{3}$, etc. The collection of weights so obtained is often called the *tree measure* of the sample space described by the tree diagram. Note that this scheme has assigned to the four points of S the same weights as those obtained by our first analysis.

The branch probabilities in the second stage are conditional probabilities. For example, $\frac{1}{3}$ is the probability of drawing a red ball, given that urn 1 is the urn first selected.

Problem 2. From a box containing 2 red discs and 2 white discs, we draw discs one by one until two red discs have been withdrawn. What is the probability that 3 draws are required?

A tree diagram for this experiment is presented in Fig. 8 below. The required probability is $P(A)$, where $A = \{RWR, WRR\}$; that is, $P(A) = \frac{1}{6} + \frac{1}{6} = \frac{1}{3}$.

In summary, weights are assigned to a sample space built sequentially in terms of stages in a tree diagram in the following way. The sum of the

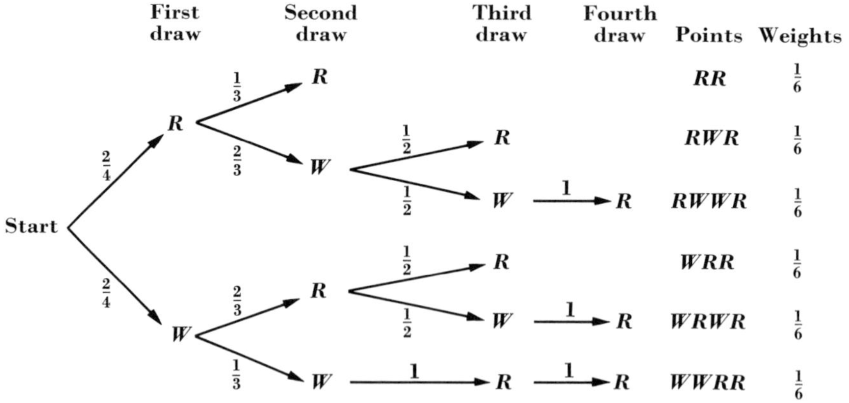

Figure 8

probabilities assigned to the branches emanating from each point is always 1. Each branch probability is the conditional probability of the outcome at the end of the branch, given the succession of outcomes on the path leading to the branch point. Finally, each path of the tree diagram represents a point in the sample space, and the weight assigned to such a point is the product of all the branch probabilities of the path.

The technique of building up an assignment of weights for a sample space in stages as illustrated in Problems 1 and 2 is very useful in dealing with experiments that involve a sequence of actions. It can be shown that the tree measures obtained by multiplying the branch probabilities always lead to a valid set of weights for the points of the resulting sample space—that is, a set of weights with sum 1.

The next problem has appeared in many of the treatises on probability theory.

PROBLEM 3. *Birthday problem.* A clerk registering voters is struck by the fact that in a group of 20 or 30 people, he frequently finds at least two people who have the same birthday. What is the probability of this coincidence in a group of n people?

The experiment of asking each of n people in turn his birthday may be pictured as an n-stage tree diagram in which the first stage results in the announced birthday of the first person asked, the next stage the birthday given by the second person, etc. Assuming the independence of successive stages of this experiment (in particular, then, no twins in the group), there are 365^n paths in this tree, to each of which we assign a weight $1/365^n$. If A is the event "at least two birthdays the same," then the complementary event in this space S is the event $\overline{A}$: "all n birthdays different." There are

$$n(\overline{A}) = 365 \cdot 364 \cdots (365 - n + 1)$$

paths in the event $\overline{A}$. Hence

$$P(\overline{A}) = \frac{n(\overline{A})}{365^n} = \frac{365 \cdot 364 \cdots (365 - n + 1)}{365^n}$$

and

$$P(A) = 1 - P(\overline{A}).$$

It is not easy to compute $P(\overline{A})$ if n is large. Of course, $P(\overline{A})$ can be approximated by means of logarithms for an n which is not too large. It may thus be shown that

$$P(\overline{A}) \doteq 0.493 \qquad \text{if} \qquad n = 23$$

so that

$$P(A) \doteq 0.507 \qquad \text{if} \qquad n = 23,$$

and furthermore, $P(A) > 0.507$ for $n > 23$. In other words, in a group of 23 people, the probability of at least two people having the same birthday is greater than $\frac{1}{2}$, so that the clerk's observations are corroborated by the theory of probability. Incidentally, if a group contains 60 people, the chances that at least two have the same birthday are greater than $\frac{99}{100}$.

The final problem of this section illustrates what is sometimes referred to as "probability by hindsight." Such situations lead to theorems known in probability theory as "Bayes's theorems for computing *a posteriori* probabilities."

PROBLEM 4. Sunshine, a race horse, prefers to run in good weather. When it rains he has a 50–50 chance of winning, but in fair weather the probability he will win is $\frac{3}{4}$. In Mudville, where he races, it is known that in June one-third of the days are rainy, the rest fair. If you read in the paper that Sunshine won his race on June 25, what is the probability that Mudville had a rainy day on that date?

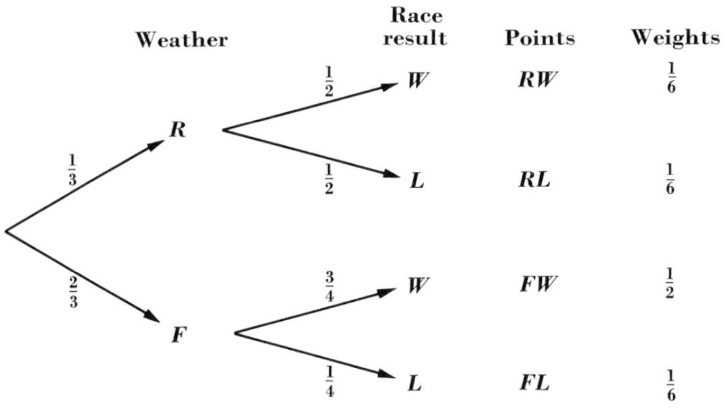

FIGURE 9

The points of our sample space can be shown as paths of a tree with two branching stages, the first stage giving the weather in Mudville on June 25, the second stage giving the result of Sunshine's race on that day. The various branch probabilities are shown in the tree diagram of Fig. 9. From this tree diagram, we see that the sample space is given by

$$S = \{RW, RL, FW, FL\}.$$

The two events A: "win" and B: "rainy day" are given by

$$A = \{RW, FW\} \quad \text{and} \quad B = \{RW, RL\}.$$

We are asked to find $P(B|A)$. Since $B \cap A = \{RW\}$, we have

$$P(B \mid A) = \frac{P(B \cap A)}{P(A)} = \frac{\frac{1}{6}}{\frac{1}{6} + \frac{1}{2}} = \frac{1}{4}.$$

Before reading the news item about the race, all we can say about the weather in Mudville on June 25 is that chances for rain are 1 in 3. After learning that Sunshine won his race that day, we then revise our statement of the chances of rain in Mudville to 1 in 4. The original probability $\frac{1}{3}$ is frequently called an *a priori* probability (meaning "at the start" or on the basis of past experience) and the conditional probability $\frac{1}{4}$ is called an *a posteriori* probability (meaning "after the fact" or on the basis of current experience).

EXERCISES

1. An ordinary penny is tossed. If it lands heads, a die is thrown; if the penny shows tails, the penny is thrown again. Construct the tree diagram for this sequence of experiments, assign appropriate tree measures to the points of the space, and find the probability that

(a) the die is thown and shows a 1,
(b) the penny is tossed twice.

2. An urn contains 3 red balls and 4 white ones. The following sequence of actions is performed: one ball is drawn from the urn and its color noted; it is then returned to the urn and 2 more balls of this same color are added to the urn; then a second ball is drawn from the augmented urn.

(a) Make a tree diagram for this sequence of actions, using a two-stage tree, the first-stage outcomes giving the color of the first ball drawn and the second-stage outcomes giving the color of the second ball drawn.

(b) What is the probability that both balls are red? the first ball drawn is red? the second ball drawn is red?

3. You know that two defective light bulbs are mixed in with three good ones. You start testing the bulbs one by one until you find the two defectives.

(a) Construct a tree diagram for this experiment.

(b) What is the probability that you find the two defective bulbs as the first two tested? that it takes three tests to find both defective bulbs? that it takes four tests? Would you ever have to test all five bulbs?

4. Find to two significant digits the probability that in a group of six people at least two have the same birthday.

5. (a) If you walk into a room containing five people, what is the probability that at least one of these people will have the same birthday as yours?

(b) How large a group of people is needed for the probability to be greater than $\frac{1}{2}$ that at least one of the group will have the same birthday as yours?

6. Urn 1 has 3 red balls and 2 white ones, while urn 2 has 4 red balls and 1 white ball. A ball is chosen from urn 1 and placed in urn 2; then a ball is drawn from urn 2. What is the probability that the second ball drawn is red? Given that the second ball is red, what is the probability that a red ball was transferred from urn 1 to urn 2 before this second draw?

7. *Bertrand's coin problem.* Of three identical boxes, one contains two gold coins, one contains a gold and a silver coin, and the third contains two silver coins. A box is selected at random and a coin taken from it. Given that the chosen coin is gold, what is the probability that the other coin in the selected box is also gold? (Use a two-stage tree with first-stage selection of box and second-stage choice of coin from that box.)

8. Urn 1 contains 3 red balls and 2 white ones and urn 2 has 4 red balls and 1 white ball. An urn is chosen at random and a ball drawn from this urn. If the ball drawn turns out to be red, what is the probability that urn 1 was chosen?

9. Using the urns of Exercise 8, two balls are drawn in succession out of the urn chosen at random.

(a) Make a tree diagram for this sequence of actions. (This is a three-stage tree.)

(b) If both balls drawn are red, what is the probability that urn 1 was chosen?

10. To make up the contents of an urn, two fair coins are tossed. If both coins land heads, 2 red balls are placed in the urn; if both land tails, 2 white balls are placed in the urn, and if one coin lands heads, the other tails, then 1 red ball and 1 white ball are put in the urn. A person who has not seen this preliminary experiment and hence does not know the composition of the urn, draws out one ball which happens to be red. What is the *a posteriori* probability for each of the three possible cases of the contents of the urn?

11. In Exercise 10, suppose that the person is allowed to replace the first ball drawn and then draw a second ball from the urn, and suppose that both draws show a red ball. Find the three probabilities asked for in Exercise 10.

12. One of two coins is weighted in favor of heads so that the probability of a head is four times that of a tail. The other coin is a fair coin. If one of the two coins is chosen at random and tossed twice, showing heads on both tosses, what is the probability that it is the weighted coin?

6. BINOMIAL DISTRIBUTION

We shall apply the tree measure technique in this section to an important type of experiment, one which consists of a succession of repetitive actions. Each of these actions results in one of two outcomes, traditionally labeled "success" and "failure." It is assumed that the sequence of actions or trials of the experiment form a completely independent set. We first consider an example of such an experiment.

PROBLEM 1. A coin weighted 2 to 1 in favor of heads is tossed three times in succession. What is the probability of no heads? exactly 1 head? exactly 2 heads? all 3 heads?

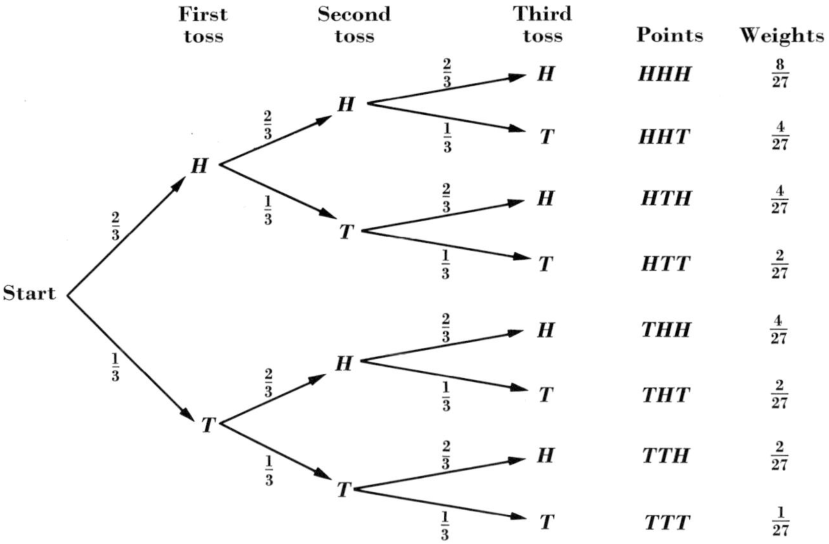

FIGURE 10

We shall build a space using a tree diagram having three stages, one for each toss of the coin (see Fig. 10).

The probabilities asked for are:

$$P\{\text{no heads}\} = P\{TTT\} = \tfrac{1}{27},$$
$$P\{1 \text{ head}\} = P\{HTT, THT, TTH\} = 3 \cdot \tfrac{2}{27} = \tfrac{2}{9},$$
$$P\{2 \text{ heads}\} = P\{HHT, HTH, THH\} = 3 \cdot \tfrac{4}{27} = \tfrac{4}{9},$$
$$P\{3 \text{ heads}\} = P\{HHH\} = \tfrac{8}{27}.$$

Labeling the outcome head as a "success" and the outcome tail as "failure," the coin tossed has probability $p = \frac{2}{3}$ of success on each toss and probability $1 - p = \frac{1}{3}$ of failure on each toss. We note that if X is used to denote the number of successes in the three trials of this experiment, we may summarize the above results as follows: X has possible values $k = 0, 1, 2, 3$, with

$$P\{X = k\} = \binom{3}{k}\left(\frac{2}{3}\right)^k\left(\frac{1}{3}\right)^{3-k}.$$

Thus, without drawing the tree of Fig. 10, we may visualize the three stages, with each arrow leading to H (success) always labeled $\frac{2}{3}$ and each arrow leading to T (failure) labeled $\frac{1}{3}$. We then reason that, for example, the case of exactly 1 success (and hence 2 failures) in the 3 trials consists of those paths each of which passes through one H and two T's and hence each of which has probability $(\frac{2}{3})(\frac{1}{3})^2$. The binomial coefficient $\binom{3}{1}$ is a convenient symbol for counting the number of these paths as the number of unit subsets of the set of stages $\{1, 2, 3\}$ which indicate the particular stage at which H occurred. Similarly, for the case of exactly 2 successes, the binomial coefficient $\binom{3}{2}$ counts the number of 2-subsets from the set $\{1, 2, 3\}$, each such 2-subset denoting the two stages at which success occurred. Since each of these paths has probability $(\frac{2}{3})^2(\frac{1}{3})$, we have $\binom{3}{2}(\frac{2}{3})^2(\frac{1}{3})$ as the probability of 2 successes.

The general argument for arriving at a set of probabilities for 0, 1, 2, . . . , n successes in n independent trials of an experiment having success probability p and failure probability $1 - p$ on each trial is exactly like that of our illustration in Problem 1. The event consisting of exactly k successes out of n independent trials is made up of those paths in an n-stage tree which pass through k S's and $n - k$ F's. Since the trials are independent with success probability p on each trial, each branch leading to an S has probability p and each branch leading to an F has probability $1 - p$. That is, the tree measure for any one of the paths in this event is $p^k(1 - p)^{n-k}$ and there are $\binom{n}{k}$ such paths. Thus, letting X be the number of successes, we have

$$P\{X = k\} = \binom{n}{k} p^k(1 - p)^{k-n}, \qquad k = 0, 1, 2, \ldots, n.$$

Note that this is a formula for a set of $n + 1$ probabilities, one for each possible number of successes in the n trials. This set of probabilities is called a *set of binomial probabilities*, or a *binomial distribution*, because of the following considerations.

If we expand the binomial $[(1 - p) + p]^n$, writing the coefficients in terms of the symbols called binomial coefficients (see Appendix 1), we have

$$1 = [(1 - p) + p]^n = \binom{n}{0}(1 - p)^n + \binom{n}{1}(1 - p)^{n-1} \cdot p$$
$$+ \binom{n}{2}(1 - p)^{n-2} \cdot p^2 + \cdots + \binom{n}{n-1}(1 - p) \cdot p^{n-1}$$
$$+ \binom{n}{n} p^n.$$

Note that the terms in the sum on the right are just the binomial probabilities for 0, 1, 2, . . . , $n - 1$, n successes in the repeated trial experiment discussed above and, of course, the total of these probabilities is 1.

We shall illustrate the use of the binomial distribution in the following problems.

Problem 2. A fair coin is tossed five times. Find the probability of the events A: "exactly one head" and B: "at least one head."

Letting X be the number of heads in the five tosses, we have

$$P(A) = P\{X = 1\} = \binom{5}{1}\left(\frac{1}{2}\right)\left(\frac{1}{2}\right)^4 = \frac{5}{32}.$$

The event B is the union of the events described by $\{X = 1\}$, $\{X = 2\}$, $\{X = 3\}$, $\{X = 4\}$, $\{X = 5\}$ or, equivalently, it is the complement of the event $\{X = 0\}$; so

$$P(B) = 1 - P\{X = 0\} = 1 - \binom{5}{0}\left(\frac{1}{2}\right)^0\left(\frac{1}{2}\right)^5 = 1 - \frac{1}{32} = \frac{31}{32}.$$

Problem 3. Find the probability of obtaining more than 2 but fewer than 6 successes in 12 independent trials of an experiment with success probability $\frac{1}{3}$ on each trial.

If X is the number of successes in the 12 trials, we wish to find the probability of the event A which is the union of events $\{X = 3\}$, $\{X = 4\}$, $\{X = 5\}$, i.e., the sum of the binomial probabilities of 3, 4, 5 successes in the 12 trials:

$$\begin{aligned} P(A) &= \binom{12}{3}\left(\frac{1}{3}\right)^3\left(\frac{2}{3}\right)^9 + \binom{12}{4}\left(\frac{1}{3}\right)^4\left(\frac{2}{3}\right)^8 + \binom{12}{5}\left(\frac{1}{3}\right)^5\left(\frac{2}{3}\right)^7 \\ &= \frac{1331 \cdot 2^8}{3^{12}} \doteq 0.64. \end{aligned}$$

The inspection of manufactured articles might be considered a sequence of independent trials, each trial being an inspection of an article resulting in one of the two outcomes, acceptance or rejection. The probability that an article will be rejected is called the *process average fraction defective.* The following problem corrects a frequent fallacy about the meaning of process average fraction defective.

Problem 4. In a certain manufacturing process, the process average fraction defective is $\frac{1}{20}$. Is it correct to infer that there is exactly one defective in each run of 20?

Let A be the event "exactly one defective in a run of 20" and label the outcome of getting a defective as "success." Using the binomial distribution for X, the number of successes, we have

$$P(A) = P\{X = 1\} = \binom{20}{1}\left(\frac{1}{20}\right)\left(\frac{19}{20}\right)^{19} \doteq 0.3725.$$

Hence the chances are only about 3 in 8 that a run of 20 will produce exactly one defective.

PROBLEM 5. How many times must a fair coin be tossed for the chances to be 99 percent or better that at least one head shows?

The event A: "at least one head" is the complement of the event "no heads"; hence we have

$$P(A) = 1 - P\{X = 0\} = 1 - (\tfrac{1}{2})^n,$$

using the binomial distribution for a sequence of unknown number n of trials with success (head) probability $\frac{1}{2}$. We require n to be such that this probability is 0.99 or better, that is,

$$\begin{aligned} 1 - (\tfrac{1}{2})^n &\geq 0.99, \\ (\tfrac{1}{2})^n &\leq 0.01, \\ 2^n &\geq 100. \end{aligned}$$

Solving this inequality, we find that the first integer n satisfying the condition is $n = 7$. That is, the coin must be tossed seven times to ensure a probability of 99 percent or better for the event "at least one head."

EXERCISES

1. A certain team has probability $\frac{2}{3}$ of winning whenever it plays. What is the probability that the team will win

(a) exactly 4 out of 5 games? (b) at most 4 out of 5 games?
(c) exactly 4 out of 5 games, given that it has already won the first 2 of the 5-game series?

2. Over a period of time, it is found that 10 percent of the fuses produced by a certain manufacturing process are defective. Find the approximate probability that in a sample of 10 fuses selected at random from the production line, there will be

(a) no defectives, (b) at least one defective, (c) at most one defective.

3. Compare the probability of obtaining at least one six in 4 tosses of a fair die with the probability of obtaining at least one double-six in 24 tosses of two fair dice.

This problem is known as de Méré's paradox. Chevalier de Méré, a gambler, thought the two probabilities ought to be equal because, he argued, "With two dice, each possibility on one die occurs with each of six possible faces of the second die; hence the probability of at least one six in *four* tosses of a die should be the same as the probability of at least one double-six in *twenty-four* tosses of two dice."

4. How many times must a fair die be thrown to make the probability greater than $\frac{9}{10}$ that at least one six shows?

5. A baseball player's batting average is .250. What is the probability that he has exactly 2 hits in 4 times at bat? at least one hit in 4 times at bat?

6. Consider the following two experiments involving an urn with 6 white and 4 red balls.

Experiment 1. Two balls are drawn in succession without replacement from the urn and their colors noted.

Experiment 2. One ball is drawn from the urn and its color noted; it is then returned to the urn and a second ball is drawn, its color noted.

Let X be the number of white balls obtained in Experiment 1 and Y the number of white balls in the two draws of Experiment 2.

(a) What are the possible values of X? of Y?

(b) Compute the probability for each of the possible values of X. Do the same for Y and compare.

(c) Is either of these experiments properly describable as a binomial experiment?

7. APPLICATIONS

One of the objectives of the theory of probability is to develop methods for use in decision-making. The following problems illustrate this aspect of probability. While the data in each problem are much simpler than those encountered in practice, the methods of solution are quite similar to those in actual use. An excellent and very readable reference for this topic is Chapter 7 of Ref. 1.*

PROBLEM 1. The Whirligig Hardware Company sells eggbeaters to restaurants in cartons of ten. An inspector from the BBB (Better Beaters Bureau) decides to inspect a carton. He considers the following two inspection plans.

Plan 1. He will take two eggbeaters from the carton. If both are satisfactory, he will accept the carton; otherwise, he will reject it.

Plan 2. He will take two eggbeaters from the carton. If both are satisfactory, he will accept the carton; if both are defective, he will reject the carton. If one is satisfactory and one defective, he will take a third eggbeater from the carton and accept or reject the carton according as the third beater is satisfactory or defective.

Which is the better inspection plan?

If the actual number of defective beaters in the carton were known, we could compute the probabilities for the acceptance of the carton under each plan and base our preference on a comparison of these probabilities. Of course, if we knew the actual number of defectives in the carton, no inspection plan would be needed anyway!

---

* The references cited in the text are to be found on p. 42.

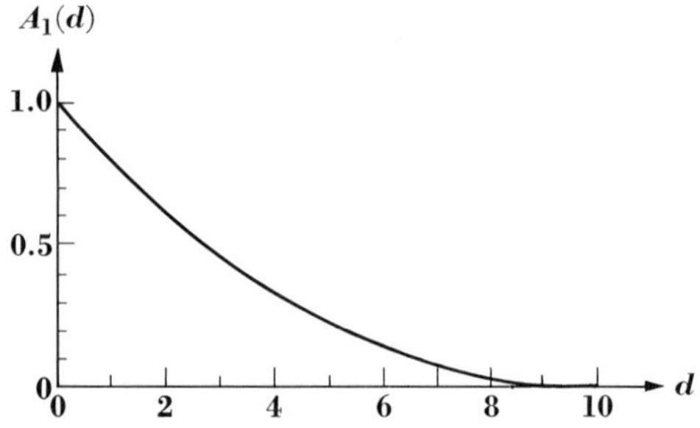

FIG. 11. Operating-characteristic curve for Plan 1.

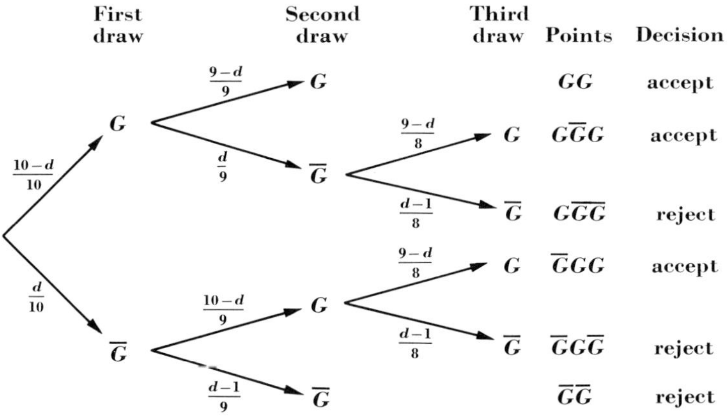

FIGURE 12

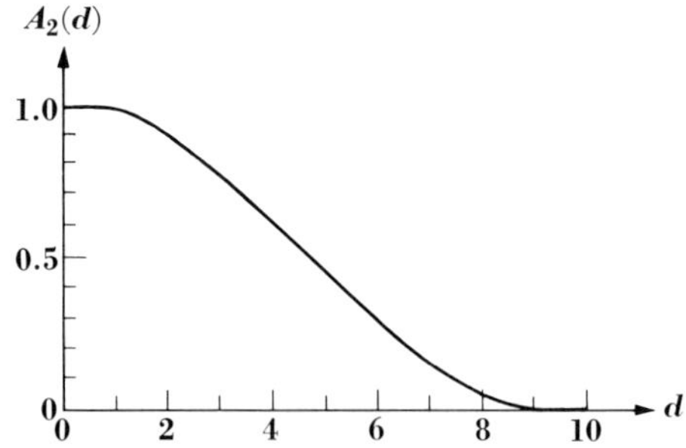

FIG. 13. Operating-characteristic curve for Plan 2.

To examine more closely the features of the two plans, let d be the unknown number of defective beaters in the carton, hence $10 - d$ the number of satisfactory ones. Let A be the event "acceptance of carton."

Plan 1 suggests the collection of all 2-subsets from the set of ten beaters in the carton as a sample space S. Event A is the collection of such 2-subsets containing two satisfactory beaters. Hence

$$P(A) = \frac{\binom{10-d}{2}}{\binom{10}{2}} = \frac{(10-d)(9-d)}{90} = \left(1 - \frac{d}{10}\right)\left(1 - \frac{d}{9}\right).$$

Note that $P(A)$ varies with d, the number of defectives in the carton and that this probability gives the probability of Plan 1's *acceptance* of the carton as a function of d. For the purposes of this problem, we rename this probability as $A_1(d)$ for the acceptance probability in terms of d under Plan 1; that is

$$A_1(d) = \left(1 - \frac{d}{10}\right)\left(1 - \frac{d}{9}\right), \qquad d = 0, 1, 2, \ldots, 10.$$

Figure 11 shows a graph of the function $A_1(d)$. It is a smooth curve connecting the plotted points $(d, A_1(d))$ for d an integer between 0 and 10 inclusive. Such a graph is called an *operating-characteristic* curve, or *OC-curve*, for the inspection plan.

To compute the acceptance probability for Plan 2, a tree diagram is useful. We shall consider the first stage to be the drawing of the first beater, the second stage selecting a second beater, and a third stage for the draw of a third beater when needed. This tree is sketched in Fig. 12.

We have used G and $\overline{G}$ to denote the outcomes "good or acceptable beater" and "defective beater," respectively, and have indicated for each arrow the appropriate probability. The event A: "acceptance of carton" consists of the three paths GG, $G\overline{G}G$, $\overline{G}GG$, and adding the tree measures for these three points, we obtain for the acceptance probability of Plan 2

$$A_2(d) = \frac{(10-d)(9-d)}{90} + \frac{(10-d)(d)(9-d)}{720} + \frac{d(10-d)(9-d)}{720}$$
$$= \left(1 - \frac{d}{10}\right)\left(1 - \frac{d}{9}\right)\left(1 + \frac{d}{4}\right).$$

In Fig. 13, we have drawn the OC-curve for Plan 2 from plotted pairs of values of $(d, A_2(d))$.

The distance from the horizontal axis to the OC-curve of either plan at an integer d gives the acceptance probability for the carton when the actual number of defectives is d. For example, we see by looking at the OC-curve for Plan 1 that this plan accepts less than one-half the time cartons

Table 1

BINOMIAL DISTRIBUTIONS FOR $n = 10$

(Entries list probabilities of exactly k successes in ten independent trials with success probability p on each trial.)

k \ p	0.10	0.25	0.50	0.75	0.90
0	0.349	0.056	0.001	0.000	0.000
1	.387	.188	.010	.000	.000
2	.194	.282	.044	.000	.000
3	.057	.250	.117	.003	.000
4	.011	.146	.205	.016	.000
5	.001	.058	.246	.058	.001
6	.000	.016	.205	.146	.011
7	.000	.003	.117	.250	.057
8	.000	.000	.044	.282	.194
9	.000	.000	.010	.188	.387
10	.000	.000	.001	.056	.349

having as few as 3 defectives. On the other hand, from the *OC*-curve for Plan 2, we see that there is an even chance that cartons having as many as 5 out of 10 defectives will be accepted. As a restaurant owner about to order these beaters, wou would probably prefer Plan 1 to Plan 2. On the other hand, from the point of view of the producer of the eggbeaters, Plan 1 might seem too stringent when its acceptance probabilities for cases of cartons with few defectives (which might be minor flaws, anyway) are compared with those of Plan 2. However, no matter which plan is used, both the consumer and the producer can tell by looking at the *OC*-curve to what extent, in the sense of probability, they are being protected or harassed by the BBB.

In industrial acceptance sampling, the process of inspecting items under mass production conditions is usually considered as a sequence of independent trials, the process average fraction defective giving the "success" probability on each trial. The *OC*-curves for such plans are constructed by means of binomial distribution tables. A section of such a table for 10 independent trials is given in Table 1. The column headings of the table list various values of the success probability p for binomial distributions with $n = 10$, and the row headings list the successive values of k, the number of successes, from 0 to 10. For example, the entry 0.146 in the column with $p = 0.25$ and row $k = 4$ gives the probability to three decimal places of exactly four successes in ten independent trials, in each of which the success probability is $p = 0.25$. This probability was computed, of course, by using the binomial probability formula

$$\binom{10}{4}(0.25)^4(0.75)^6 \doteq 0.146.$$

We shall use this table in the following problem.

Problem 2. Ten articles are selected at different times during the day at a factory producing large quantities of the product. If two or fewer defectives are found, the day's run is judged ready to sell. If more than two defectives are found, the day's run is subjected to further inspection and the machinery is overhauled. What is the probability that the day's run will be accepted when the actual process fraction defective is 0.50? when it is 0.75? when it is 0.90? What is the probability that the decision to overhaul the machinery will be made when the actual process average defective is 0.1? when it is 0.25?

The acceptance probability for a given process average fraction defective p is the sum of the probabilities of 0, 1, 2, defectives in the set of ten articles selected. Using Table 1 and denoting the acceptance probability by $A(p)$, we have

$$\begin{aligned} A(0.5) &= 0.001 + 0.010 + 0.004 = 0.055, \\ A(0.75) &= 0.000 + 0.000 + 0.000 = 0.000, \\ A(0.90) &= 0.000. \end{aligned}$$

Thus, if the process average fraction defective is 0.5 so that half the items produced are defective, there is about 1 chance in 20 that the day's run will be accepted; for process average fraction defectives of 0.75 or 0.90, the chances are negligible for acceptance of the day's run.

To find the probabilities that the day's run will be subjected to further inspection and machinery overhauled when $p = 0.10$ and when $p = 0.25$, we first compute from Table 1 the acceptance probabilities for these cases:

$$\begin{aligned} A(0.10) &= 0.349 + 0.387 + 0.194 = 0.930, \\ A(0.25) &= 0.056 + 0.188 + 0.282 = 0.526. \end{aligned}$$

Hence the probability of requiring overhaul when process average fraction defective is 0.10 is

$$1 - A(0.10) = 0.07,$$

or 7 chances in 100; when $p = 0.25$, the probability of overhaul is

$$1 - A(0.25) = 0.474,$$

almost a 50–50 chance.

EXERCISES

1. (a) For Problem 1 of this section, find the acceptance probability $A_3(d)$ for a third plan:

Plan 3. Take out two eggbeaters from the carton of ten. If both are defective, reject the carton; otherwise accept it.

(b) Graph the *OC*-curve for Plan 3 and compare it with those for Plan 1 and Plan 2.

2. Consider the following two inspection schemes for a shipment of 100 tires:

Plan 1. Test two tires from the shipment. If both are satisfactory, accept the shipment; otherwise reject it.

Plan 2. Test two tires from the shipment. If both are defective, reject the lot; otherwise accept it.

Compute formulas for the acceptance probability as a function of the number of defectives d for each plan and sketch the corresponding *OC*-curves, plotting values of d at 0, 10, 20, . . . , 100. Comment on the two plans, individually and in comparison.

3. Joe claims he can call the toss of a coin correctly more often than not. To test his claim you devise the following experiment and decision plan. You will toss the coin 10 times, each time asking Joe to state in advance the way the coin will land. If he is correct 7 or more times, you will agree with his claim; otherwise you will say he is just guessing. Using Table 1 answer the following questions:

(a) What is the probability under your decision plan that you will give Joe credit for his claim when he is actually just guessing each time?

(b) What is the probability that you accuse Joe of just guessing when actually he has a "success" probability (probability of being right in calling a toss) of $\frac{3}{4}$?

4. The standard treatment for a certain disease leads to cures in one-fourth of the cases. A new treatment is devised which is said to produce cures in three-fourths of the cases. The new treatment is tried out on 10 people having the disease. If 7 or more are cured, the new treatment will be adopted. If 3 or fewer are cured, the new treatment will be discarded as not worth further research. If the number cured is 4, 5, or 6, then judgment on the new treatment will be deferred, but studies investigating its properties will continue.

(a) Find the probability of each of these three possible actions, given that the new treatment is no more effective than the old, that is, given that $p = \frac{1}{4}$ for the ten independent trials.

(b) Find the probability of each of the actions if the new treatment is actually as effective as claimed.

5. A famous problem in the literature of mathematical statistics deals with a lady who claims to have special tea-tasting powers, namely to be right considerably more often than not in diagnosing which item was put into the cup first—the tea, or the milk, which she insists a cup of tea requires. After discussing some preliminary attempts at devising an experiment to test her claim, J. Neyman in his book, *First Course in Probability and Statistics* (Henry Holt and Co., 1950), presents the following decision procedure as being one to which

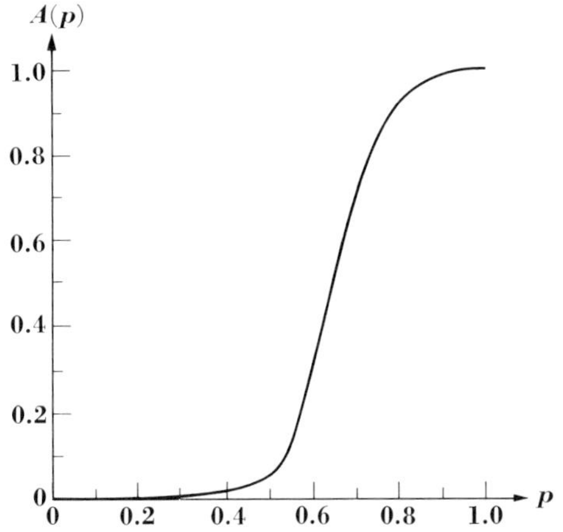

FIG. 14. Operating-characteristic curve for lady-tasting-tea experiment.

both the lady and a somewhat skeptical experimenter might agree. In a continued set of experiments, the lady tastes each of 40 pairs of cups of tea, each pair containing tea made by each of the two methods. If she correctly diagnoses 26 or more of these pairs, her claim is granted; otherwise the experimenter concludes that she is merely guessing.

The *OC*-curve for the above decision plan is sketched in Fig. 14 with $A(p)$ giving the probability of accepting the lady's claim when her success probability on each try is p. Examine this *OC*-curve and then discuss briefly why you think this plan should or should not be agreeable to both the lady and the experimenter.

8. FINITE MARKOV CHAINS

In the last section, we presented an introduction to the role of probability in decision-making. The theory of probability is also useful in formulating explanations of the underlying mechanisms of certain phenomena for the purpose of making predictions. One such example is contained in the independent trials process of Section 5. A second example is that of Finite Markov Chains, a very brief instance of which will be discussed in this section.

Since elementary notions of matrices (in particular, matrix multiplication) are used throughout this section, the reader is advised to turn first to Appendix 2 to acquire familiarity with these ideas or to refresh his memory of such matters.

PROBLEM 1. Suppose that a simplified computer produces only "0" and "1." It has the peculiar property that whenever a "0" is fed into the machine, there is a 0.9 chance of it being reproduced as "0," but that with probability 0.1 it will be switched to a "1." On the other hand, an

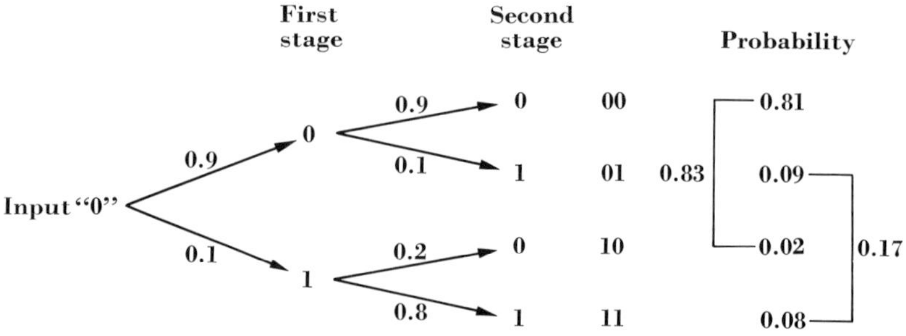

FIGURE 15

input of "1" will be reproduced as "1" with probability 0.8 and changed to "0" with probability 0.2. Suppose that the process of input "0" or "1" and output "0" or "1" is continued for two, three, four, and even very many stages. What is the probability of the final output being a "0" if the initial input was a "0"? if the initial input was a "1"?

In Fig. 15 a tree is drawn for two stages of this process, starting with input "0." Note that after two stages, the probability of final output "0" is

$$P\{00, 10 \mid \text{input "0"}\} = 0.81 + 0.02 = 0.83$$

and, similarly, the probability of final output "1" is

$$P\{01, 11 \mid \text{input "0"}\} = 0.09 + 0.08 = 0.17.$$

We leave it to the reader to draw a like tree for two stages of the process with original input "1" and to see that

$$P\{00, 10 \mid \text{input "1"}\} = 0.34, \qquad P\{10, 11 \mid \text{input "1"}\} = 0.66.$$

A convenient way of summarizing these four probabilities is in a two-by-two array, called a matrix, which we shall denote by

$$P_2 = \begin{pmatrix} 0.83 & 0.17 \\ 0.34 & 0.66 \end{pmatrix}.$$

In this array, the first row gives the probabilities of final outputs "0" and "1" starting with input "0"; the second row gives the probabilities of these outputs starting with input "1." We have called this matrix P_2 to emphasize that these are probabilities after *two* stages of the process. In this notation, we may use a matrix P_1, which we shall designate simply as P, to give the first stage probabilities from input "0" to outputs "0"

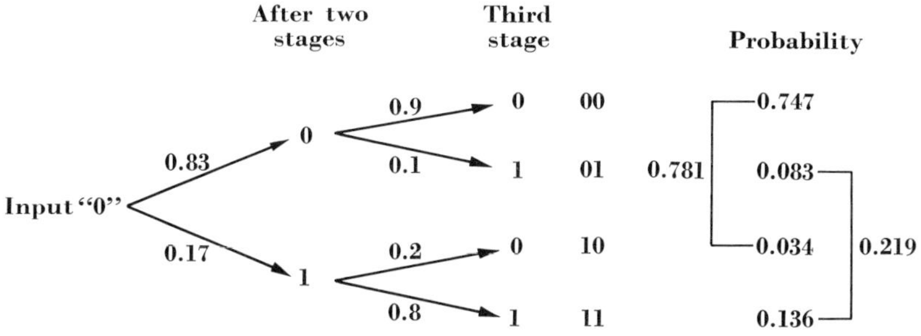

FIGURE 16

and "1," and from input "1" to these outputs. Thus

$$P = \begin{pmatrix} 0.9 & 0.1 \\ 0.2 & 0.8 \end{pmatrix}.$$

An interesting relationship exists between the matrix P and the matrix P_2 (cf. Appendix 2). The reader may verify that

$$P_2 = \begin{pmatrix} 0.9 & 0.1 \\ 0.2 & 0.8 \end{pmatrix}^2 = P^2.$$

That is, the matrix of probabilities of outputs "0" and "1" after *two* stages, starting with initial input of "0" for the first row and "1" for the second row, is the square of the matrix P of first-stage probabilities.

Suppose that we now go on to the third stage of this process. We could extend the tree of Fig. 15 to include one more stage (and, similarly, the tree corresponding to that of Fig. 15 in the case of original input "1"). Instead, we condense this tree as indicated in Fig. 16, making use of the computations already made of the probabilities after two stages.

We have written on the first branches from initial input "0" to outcomes "0" and "1" the numbers 0.83 and 0.17, already computed for the probabilities of these outputs after two stages. Again, we ask the reader to make a similar tree for original input "1" to check that the probabilities of outputs "0" and "1" after three stages are then 0.438 and 0.562. Thus

$$P_3 = \begin{pmatrix} 0.781 & 0.219 \\ 0.438 & 0.562 \end{pmatrix}$$

is the matrix of probabilities of outputs "0" and "1" after three stages, the first row giving these probabilities when initial input was "0" and the second row in the case of initial input "1."

Furthermore, it can be checked that

$$P_3 = P^2 \cdot P = P^3.$$

In general, it can be shown that the matrix P_n of probabilities after n stages of the process is given by the nth power of the matrix P; that is,

$$P_n = P^n.$$

At this point it is convenient to introduce some terminology for this type of process.

Although there are two outcomes, "0" and "1," at each stage, the above experiment is *not* an example of a binomial experiment or independent trials process of Section 5, since the probability of arriving at a "0" at any stage depends on the outcome of the preceding stage. If the preceding outcome was "0," then the probability of a "0" next is 0.9; if the preceding outcome was "1," then the probability of a "0" on the next stage is only 0.2. Similarly, the probability of a "1" depends on whether the preceding outcome was a "0" or "1." However, we need only know what the immediately preceding result was in order to know what probability to attach to each of the outcomes "0" and "1," and we always attach the same numbers (0.9 for "0" in case of an immediately preceding "0" and 0.2 for "0" in case of an immediately preceding "1," etc.). Such a sequence of trials or process is called a *Markov Chain*, and the possible outcomes at each stage are called *states*. The above example is a Markov chain with two states, "0" and "1." One can have a Markov chain with more than two states. In our example, a machine producing digits "0," "1," and "2" would have three states. If the number of states is finite, the process is termed a *finite Markov chain*.

The matrix of probabilities P at the first stage is called the *transition matrix* of the chain. In our case,

$$P = \begin{pmatrix} 0.9 & 0.1 \\ 0.2 & 0.8 \end{pmatrix}.$$

Note that to get from the outcomes after any number of stages to the two states "0" and "1" *in one step of the process*, the transition matrix P gives the branch probabilities: the first row gives the probabilities to "0" and "1" from outcome "0," the second row gives these two probabilities from outcome "1."

What we have illustrated in Problem 1 is a case of the following:

Let P be the transition matrix of a finite Markov chain having k states, $a_1, a_2, \ldots, a_k$. Then P^n is the matrix of probabilities of arriving at each of the states after n stages of the chain, the first row giving these probabilities starting in state a_1, the second row starting in a_2, etc., the last row giving the probabilities starting in a_k.

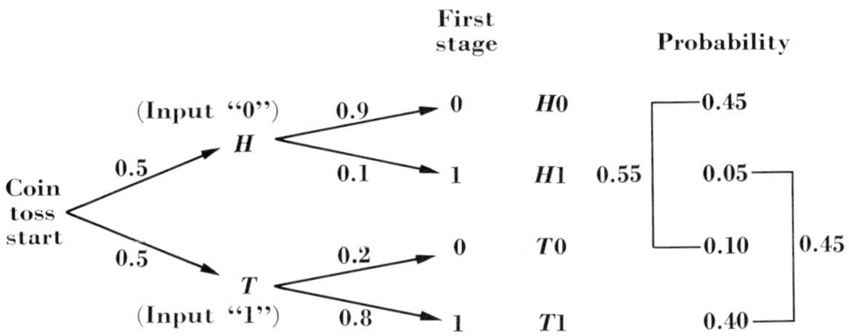

FIGURE 17

We now consider a slight variation on Problem 1.

PROBLEM 2. Suppose that in order to get the process of Problem 1 started, a person first tosses a fair coin; if the coin shows heads, he starts the process with input "0," if it shows tails, he starts the process with a "1." Find the probability of an output "0" after 1, 2, 3, . . . , n stages of the process; of an output "1" after several stages.

The tree of Fig. 17 illustrates the experiment of Problem 2. Note that in this experiment with a random start, one tree is sufficient for the determination of input. We have bracketed the probabilities of the two points $H0$, $T0$, which give outcome "0" after one stage of this process and have similarly bracketed the probabilities of the two points $H1$ and $T1$. Thus

$$P\{\text{“0”} \mid \text{coin toss start}\} = 0.55,$$
$$P\{\text{“1”} \mid \text{coin toss start}\} = 0.45.$$

We will write these two probabilities as a vector, with the understanding that the first component is the probability of output "0" and the second component the probability of outcome "1." We denote this vector by

$$X_1 = (0.55, 0.45),$$

and call it the vector of probabilities after one stage with initial start of the process by the coin toss. In this notation it is convenient to write the vector representing the coin toss probabilities as

$$X_0 = (0.5, 0.5),$$

and call this the *initial probability vector* of the process.

Suppose we now let the process continue one more stage. Again, instead of expanding the tree of Fig. 17, we may condense this as in Fig. 18 and obtain

$$X_2 = (0.585, 0.415)$$

as the probability vector for outputs "0" and "1" after two stages of the process under initial conditions of the coin toss start.

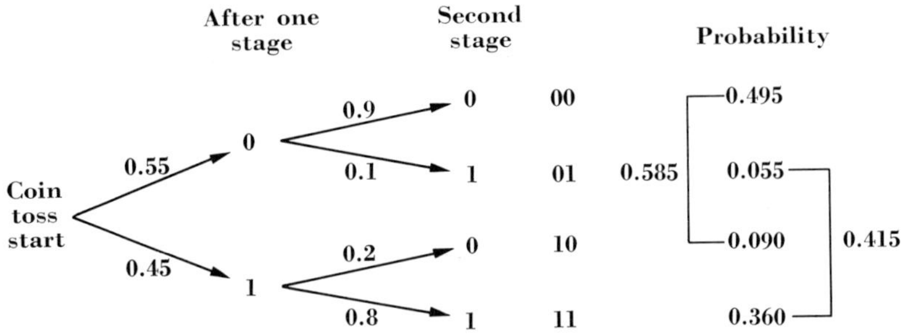

FIGURE 18

We ask the reader to verify that

$$X_2 = X_1P \qquad \text{and} \qquad X_1 = X_0P$$

so that

$$X_2 = (X_0P)P = X_0P^2.$$

That is, to obtain X_1 we can simply multiply the initial probability vector X_0 by the transition matrix P; to obtain X_2, we multiply X_0 by P^2, etc. In general,

$$X_n = X_{n-1}P$$

so that

$$X_n = X_0P^n,$$

where X_n gives the probabilities of the outcomes after n steps, starting with initial probability vector X_0.

PROBLEM 3. Is there an initial probability vector Y_0 for which $Y_0 = Y_0P$ for the transition matrix P of Problems 1 and 2?

This question may seem rather irrelevant to the questions of Problems 1 and 2, but let us consider first the consequences of a "yes" answer to the question of Problem 3. In view of the results established in the discussion of Problem 2, if there were an initial probability vector Y_0 satisfying the conditions of the present problem, we would have

$$Y_1 = Y_0P = Y_0.$$

That is, the vector of probabilities of the outcomes "0" and "1" after one stage of the process is just the same as the initial vector Y_0 which started the process in motion. But then

$$Y_2 = Y_1P = Y_0P = Y_0$$

so that Y_2 is also the same as Y_0. In fact,

$$Y_n = Y_0, \quad \textit{whatever the number of stages } n.$$

Therefore the existence of a vector Y_0 with the property $Y_0 = Y_0P$ would furnish a random device for starting the process in such a way that after any number of stages the probabilities of "0" and "1" outputs would remain the same as those given by Y_0.

Let us try a purely algebraic approach to the solution of Problem 3. We are looking for a vector

$$Y_0 = (y, 1 - y)$$

(since the two components are probabilities whose sum is 1), for which

$$(y, 1 - y)P = (y, 1 - y)$$

or

$$(y, 1 - y)\begin{pmatrix} 0.9 & 0.1 \\ 0.2 & 0.8 \end{pmatrix} = (y, 1 - y).$$

Performing the indicated multiplication, we obtain

$$(0.9y + 0.2 - 0.2y, 0.1y + 0.8 - 0.8y) = (y, 1 - y),$$
$$(0.7y + 0.2, -0.7y + 0.8) = (y, 1 - y).$$

Hence, we have

$$0.7y + 0.2 = y \qquad \text{and} \qquad -0.7y + 0.8 = 1 - y$$

or

$$y = \tfrac{2}{3} \qquad \text{and} \qquad 1 - y = \tfrac{1}{3}.$$

That is, the vector $Y_0 = (\frac{2}{3}, \frac{1}{3})$ satisfies the requirement of Problem 3.

We remind the reader that the above solution for Y_0 has the following impact in terms of our computer producing "0" and "1": If one first uses a random device to produce initial probabilities $\frac{2}{3}$ and $\frac{1}{3}$ (e.g., throw a fair die; if the die lands 1, 2, 3, 4, use input "0"; otherwise use "input 1" in machine), then under the given assumptions on the way this machine operates, after one, two, three, or any number of stages, the probability that the output of the machine is a "0" will be $\frac{2}{3}$ and the probability of output "1" will be $\frac{1}{3}$.

The connection between the answer to Problem 3 and the questions raised in Problems 1 and 2 is given in the theorem stated below after a preliminary definition.

Definition. If P is the transition matrix of a finite Markov chain and Y is a probability vector having the property

$$YP = Y,$$

then Y is called a *fixed probability vector* of the matrix P.

Theorem 4. If P is the transition matrix of a finite Markov chain having the property that some power of P has no zero entries, then

(a) P has a unique fixed probability vector Y, and Y has no zero components;

(b) as n increases, P^n approaches the matrix A which has as each of its rows the vector Y;

(c) if X_0 is any initial probability vector, then as n increases the vector $X_n = X_0P^n$ approaches the vector Y.

Note. To say that P^n *approaches* the matrix A is to say that each entry of P^n approaches the corresponding entry of A. Similarly, to say that X_n *approaches* the vector Y means that each component of X_n approaches the corresponding component of Y.

For a discussion of this theorem and other results in the theory of finite Markov chains, the reader should consult Ref. 3.

We now apply parts (b) and (c) of Theorem 4 to Problems 1 and 2 of this section.

By (b), we find in Problem 1 that P^n approaches the matrix A given by

$$A = \begin{pmatrix} \frac{2}{3} & \frac{1}{3} \\ \frac{2}{3} & \frac{1}{3} \end{pmatrix},$$

since the fixed probability vector Y of P was found in Problem 3 to be

$$Y = (\tfrac{2}{3}, \tfrac{1}{3}).$$

This result implies that if one starts with input "0," then after a sufficiently large number of steps the probabilities of outputs "0" and "1" are approximately $\frac{2}{3}$ and $\frac{1}{3}$, respectively. But if one starts with input "1," after enough steps the probabilities of outputs "0" and "1" are also approximately $\frac{2}{3}$, $\frac{1}{3}$, respectively. That is, it doesn't matter, in the long run, which input is used—a "0" will turn out with probability $\frac{2}{3}$ and a "1" will be issued with probability $\frac{1}{3}$, if the process has been in effect long enough.

Part (c) of Theorem 4, applied to Problem 2, implies that if one first flips a fair coin to get the process started, it is still the case that in the long run a "0" output will occur with $\frac{2}{3}$ probability and a "1" output with probability $\frac{1}{3}$. Furthermore (see Exercise 1 below), if one used a different random device to get the process started, the results cited above would still hold.

This discussion constitutes a mere introduction to the subject of this section. The interested reader is encouraged to consult the reference cited above for further applications of the Markov chain model in such diverse fields as genetics, sociology, economics, psychology, and physics.

EXERCISES

1. Using the transition matrix P of Problems 1, 2, and 3 of this section, consider questions of Problem 2 when, instead of tossing a fair coin, a fair die is thrown with input "0" used if a six shows, and otherwise input "1" starts the process. Find X_1 and X_2, the probability vectors giving the probabilities of outputs "0" and "1" after one stage and after two stages of the process with this start.

2. Although it is a tedious business, compute P^4, P^8, and P^{16} for the matrix P of Problem 1. Round off entries of each matrix to three decimal places.

Referring to the statement (b) of Theorem 4, your results in this exercise should give some indication of the fact that the approach of P^n to the matrix A, where here A is

$$A = \begin{pmatrix} \frac{2}{3} & \frac{1}{3} \\ \frac{2}{3} & \frac{1}{3} \end{pmatrix},$$

does not require exceptionally large n for a "close" approach.

3. For the initial probability vector of Problem 2 of this section, that is, $X_0 = (0.5, 0.5)$, compute $X_4 = X_0P^4$, $X_8 = X_0P^8$, and $X_{16} = X_0P^{16}$.

4. Compute X_4, X_8, and X_{16} for the initial probability vector used in Exercise 1 above, and compare your results with those of Exercise 3.

5. In a T-maze a laboratory animal is given the choice of going to the left and getting food or going to the right and receiving a mild electric shock. Having received food on any run, the probabilities of the animal going left or right on the next run are 0.8 and 0.2, respectively. Having received the electric shock on any run, the probabilities of it going left or right on the next run are 0.6 and 0.4, respectively.

(a) Write the matrix P of transition probabilities for this Markov chain.

(b) Find the unique fixed probability vector for P.

(c) Assuming that before any conditioning, i.e., before the first run, animals are equally likely to go to the left or to the right in the maze, what is the probability that an animal put in the maze will turn left after the first run? after the second run? after the third run? After a large number of runs, what proportion of the time, approximately, will an animal turn left?

6. This exercise has to do with people who turn in their cars for new ones each year. We shall consider only three makes of car, which for the sake of impartiality we shall call makes A, B, C. People who own make A any year tend to switch to make B with probability 0.4 and to switch to C with probability 0.2, hence to buy another car of make A with probability 0.4. Owners of make B switch to A with probability 0.2, to C with probability 0.2, hence keep make B with probability 0.6. Owners of make C have probability 0.2 of a switch to A, probability 0.4 of a switch to B, and chance of 0.4 to keep make C.

(a) Give the 3×3 matrix P of transition probabilities for this three-state chain.

(b) Find the fixed probability vector Y for P. [That is, solve the equation $(y, z, 1 - y - z)P = (y, z, 1 - y - z)$ for y and z.]

(c) If Mr. Jones has a car of make A in 1966, what is the probability that he will be driving a make A car in 1968? that he will be driving make B? make C?

(d) If Mr. Jones has make B in 1966, answer the three questions of part (c).

(e) If he has make C in 1966, answer the three questions of part (c).

(f) Whatever make Mr. Jones has in 1966, what is the long-run probability of his driving make A? make B? make C?

REFERENCES FOR FURTHER READING

1. Commission on Mathematics, C.E.E.B., *Introductory Probability and Statistical Inference for Secondary Schools*, 427 W. 117 St., New York 27, N.Y., 1959.

2. Goldberg, Samuel, *Probability; an Introduction.* Englewood Cliffs, N.J.: Prentice-Hall, 1960.

3. Kemeny, J. G., J. L. Snell, and G. L. Thompson, *Introduction to Finite Mathematics.* Englewood Cliffs, N.J.: Prentice-Hall, 1957.

4. Mosteller, F., R. E. K., Rourke, and G. B. Thomas, *Probability, a First Course.* Reading, Mass.: Addison-Wesley, 1961.

5. Neyman, J., *First Course in Probability and Statistics.* New York: Henry Holt and Co., 1950.

6. Wallis, W. A., and H. V. Roberts, *Statistics, a New Approach,* Chap. 16. Glencoe, Ill.: The Free Press, 1956.

APPENDIX 1 | SOME COUNTING PRINCIPLES

In general, counting the number of objects in large collections is a matter of devising a systematic procedure for listing the members of the collection. In this process of listing objects one notes certain patterns which enable one to arrive at the total number of objects in the collection without actually completing the listing. Some of these patterns occur frequently enough to be formulated in terms of general rules or principles for obtaining the count of such collections. These we shall summarize in this appendix.

A particularly simple principle for counting the number of possible results for performing a number of actions in sequence is the following:

> **Sequential counting principle** (SCP). Given that two or more actions are performed in a definite order, that the first action produces m possible results, for *each* of these results, the second action produces n possible results, and for *each* of these results of the first two actions, the third action produces p results, etc., then the number of possible results of this sequence of actions is the product of the numbers m, n, p, etc.

We shall illustrate this principle with an example.

PROBLEM 1. From a set of four discs numbered 1, 2, 3, 4, three discs are selected in turn and placed in a row. How many different three-digit numbers can be so formed? In Fig. 19, we have indicated schematically one way of answering this question. Start with each of the four possible discs as a first digit, then put next to it as a second digit each of the remaining discs, and for each of these second disc placements use as third digit all the remaining discs after the first two are in position. Note that the total number of three-digit numbers so formed is 24, which is the number of paths in the tree. The number of paths is the product of the number of branches in the first stage, the number of branches drawn from any one of the first-stage endpoints to the second-stage possibilities, and the number of branches from each branch point of the second stage to the third-stage endpoints; that is, $4 \cdot 3 \cdot 2$ or 24. The counting principle given above as SCP merely formulates the general result of counting the paths of a tree whose stages represent in turn the sequence of actions being performed.

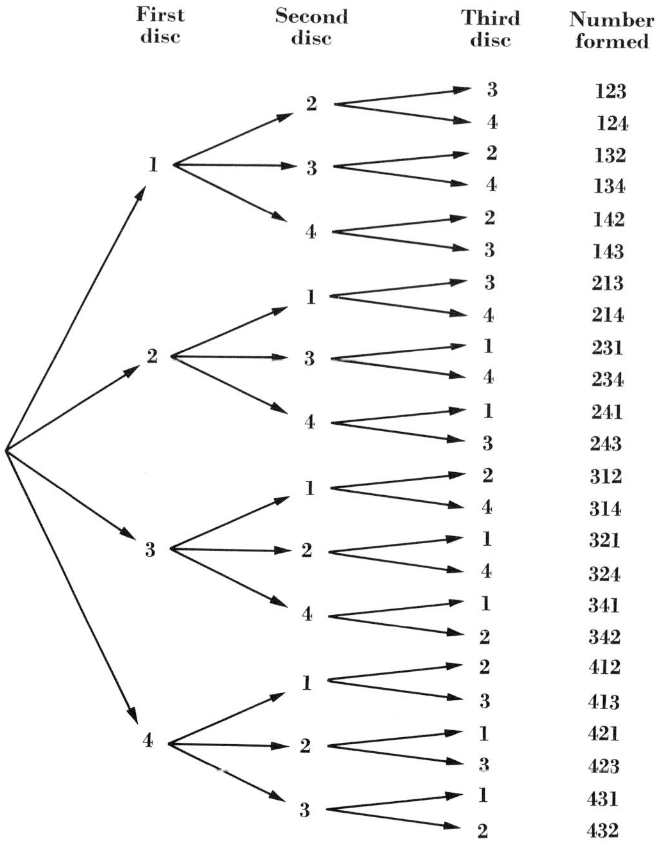

FIGURE 19

Some further illustrations show the scope of this basic counting principle.

PROBLEM 2. Find the number of possible three-digit numbers which contain only the digits 1, 2, 3, 4.

Problem 2 differs from Problem 1 in that repetitions of digits are allowed, for example in forming the numbers 233 or 333. The tree diagram to be visualized is one in which at each branching point there are four branches representing the four possible digit placements for the first, second, and third digit positions. Therefore the count is $4 \cdot 4 \cdot 4$ or 64 numbers.

In case certain of the actions performed in sequence are restricted or special in some way, it is best to provide for these particulars first, as in the following example.

PROBLEM 3. Using only the digits 1, 2, 3, 4, and not allowing repetitions, how many three-digit numbers may be formed in which the middle digit is odd?

The count is $2 \cdot 3 \cdot 2 = 12$, or the number of paths in a tree diagram with branching stages at actions representing, in sequence, the placing in position of the middle digit, then the first, and finally the last digit of the number. There are two branches at the first stage leading to each of the two possible odd digits, then three from each of these to the numbers remaining to be used for the first digit, and from each of these, two branches to the numbers left as possibilities for the third digit.

A second general counting principle may be illustrated in terms of the previous examples involving the formations of numbers as follows.

PROBLEM 4. Using the digits 1, 2, 3, 4, how many numbers can be formed if repetitions of digits are not allowed?

In this example, clearly any number k of digits, for $k = 1, 2, 3, 4$, may be used. For $k = 1$, there are 4 numbers; for $k = 2$, we obtain by SCP a total of $4 \cdot 3 = 12$ numbers; similarly, for $k = 3$, there are $4 \cdot 3 \cdot 2 = 24$; and for $k = 4$, the total is $4 \cdot 3 \cdot 2 \cdot 1 = 24$. The total count, then, is $4 + 12 + 24 + 24 = 64$.

> **Counting principle for cases of alternatives** (ACP). If an action can be analyzed into several *mutually exclusive cases*, the total number of possible results is the *sum* of the numbers giving the totals for each of the separate cases.

In Problem 4, there were four mutually exclusive cases according to the number of digits in the number formed. The totals for each of these separate cases were obtained by SCP and then these totals combined under ACP to yield the final result.

The two counting principles, SCP and ACP, are essentially sufficient for the purpose of determining the number of outcomes in a finite set of possibilities. However, it is convenient to classify certain typical counting situations which occur frequently. We shall illustrate two important cases using for an underlying set of objects the set $U = \{a, b, c, d\}$.

PROBLEM 5. (a) How many arrangements (ordered strings of symbols) are there, using three of the four symbols from U and not allowing repetitions? (b) How many subsets of three elements each can be formed from the elements in U?

The answer to the question of part (a) is $4 \cdot 3 \cdot 2$ or 24 by SCP. In the list of 24 arrangements counted in part (a), each different subset of three of the objects must occur and, furthermore, will appear six times in the list, since placing the three objects in sequence will by SCP yield $3 \cdot 2 \cdot 1$

or 6 arrangements. Hence there is a total of $\frac{24}{6}$ or 4 different *subsets* each involving three of the four symbols of U, which answers the question of part (b).

Before presenting formulas for obtaining these counts in terms of the two situations, of which the above were typical, we shall introduce some simplifying notation and some useful terminology.

Definition 1. For a positive integer n, *n factorial*, denoted by $n!$, is defined by

$$n! = n(n - 1)(n - 2) \cdots 2 \cdot 1.$$

For convenience, we also define $0!$ to be the number 1.

Definition 2. An arrangement of r symbols, not allowing repetitions among the symbols, is called a *permutation* of these r symbols. A subset of r elements is called an *r-subset* or a *combination* of r elements.

Formula 1. *Permutations.* The number ${}_nP_r$ of permutations of r symbols chosen from a set of n symbols is given by

$${}_nP_r = n(n - 1)(n - 2) \cdots (n - r + 1).$$

Formula 2. *Combinations.* The number ${}_nC_r$ or $\binom{n}{r}$ of r-subsets or combinations of r objects chosen from a set of n objects is given by

$${}_nC_r = \binom{n}{r} = \frac{{}_nP_r}{r!} = \frac{n!}{r!(n - r)!}.$$

In particular, using the definition of $0!$, we may include ${}_nC_0$ or $\binom{n}{0}$ to yield the number 1.

These two formulas are established by arguments analogous to those of parts (a) and (b) of Problem 5.

The symbols $\binom{n}{r}$ for counting combinations are often called *binomial coefficients* from their use in expanding a binomial to a positive integral power n. Thus we may write

$$(a + b)^n = \binom{n}{0} a^n + \binom{n}{1} a^{n-1}b + \binom{n}{2} a^{n-2}b^2 + \cdots + \binom{n}{n-2} a^2b^{n-2} + \binom{n}{n-1} ab^{n-1} + \binom{n}{n} b^n,$$

where, for example, the coefficient of the term $a^{n-r}b^r$ is $\binom{n}{r}$, since this is the number of r-subsets of the set $\{1, 2, 3, \ldots, n\}$ of numbered positions of the n factors $(a + b)$. Any one of these r-subsets represents the particular combination of factors from which the b term was selected. The a's are chosen from the remaining $n - r$ factors to form the particular product $a^{n-r}b^r$ in the expansion of $(a + b)^n$ by repeated use of the distributive property of numbers.

As stated earlier, the two formulas given above are convenient *when* the counting problem can be identified as of one of the types to which they apply. A safe procedure is to revert to the two basic counting principles when in doubt and to arrive at the count directly by an actual (conceptual) listing of outcomes.

EXERCISES

1. If you are given a set of five discs, numbered 1, 3, 5, 6, 9,

(a) how many two-digit numbers can be represented by placing two discs side by side?

(b) how many four-digit numbers can be represented by placing four discs in a row?

(c) how many five-digit numbers can be represented by placing five discs in a row?

2. There are exactly four different roads from town A to town B and three from town B to town C. There are also two roads from A to C bypassing town B.

(a) How many different routes are there from A to C via B?

(b) How many different routes are there from A to C?

(c) How many different round trips are there from A to C and back again to A?

3. (a) How many different individuals can there be, each with two initials and no two with the same ordered pair of initials?

(b) How many different members can there be in an organization if each person has three initials and no two persons have the same ordered triple of initials?

(c) How many different monograms should an embroiderer prepare to be ready for any person with two or three initials?

(d) If in a town of 20,000 people, each person has either two or three initials, how can you be sure that not every person has an ordered set of initials different from that of everyone else?

4. Using the set $U = \{a, b, c, d\}$ of Problem 5, list the $\binom{4}{3} = 4$ different 3-subsets of U. For each of these subsets write the $3! = 6$ different permutations of the elements of the subset. Compare the resulting list of $4 \cdot 6 = 24$ permutations of three objects from the set U with a list formed by using a tree diagram to obtain the set of 24 arrangements of three symbols from U.

5. Evaluate:

(a) ${}_{10}P_3$ (b) $\dfrac{{}_9P_4}{{}_4P_4}$ (c) ${}_9C_4$

(d) ${}_9C_5$ (e) ${}_{1000}C_{999}$ (f) ${}_{20}C_{20}$

(g) ${}_{10}P_{10}$ (h) ${}_{16}P_1$ (i) ${}_{16}C_1$

(j) $\binom{2000}{1998}$

(k) $\binom{8}{2}\binom{6}{2}\binom{4}{2}\binom{2}{2}$

(l) ${}_4P_3 + {}_4P_2 + {}_4P_1$

(m) $\binom{6}{2} + \binom{6}{3}$

6. (a) How many different committees of four can be formed from a group of eight students?
(b) How many of these contain one particular student A?
(c) How many of these committees include A and exclude another student B?
(d) How many include both A and B?
(e) How many exclude both A and B?

7. Out of a group of ten boys and seven girls, three boys and two girls are to be selected to represent a school. In how many different ways can the selection be made?

8. A certain lot consisting of ten articles has three defective and seven non-defective articles. (a) How many possible subsets of two articles can be chosen from this lot? (b) What proportion of these subsets contain two defectives? (exactly) one defective? no defectives?

9. In Exercise 8, suppose the two articles are chosen in succession from the lot of ten articles, the first article being replaced before the second is drawn. (a) How many possible choices of the two articles are there? (b) What proportion of these choices contain two defectives? one defective? no defectives?

10. By evaluating the binomial coefficients, show that, for positive integers n, r with $r \leq n$,

$$\binom{n}{r} + \binom{n}{r-1} = \binom{n+1}{r}.$$

APPENDIX 2 | MULTIPLICATION OF MATRICES

It is sometimes convenient to work with a collection of sets of numbers instead of with the individual numbers themselves. An example might be in listing the unit prices of, say, five commodities such as oranges, lemons, pears, bananas, and peaches at a particular store A. Suppose these prices are given respectively as 5¢, 3¢, 6¢, 4¢, 5¢. These numbers may be written in a *row vector*, which we call A, as

$$A = (5, 3, 6, 4, 5).$$

Similarly, if a set of prices for these same five articles at store B is given, we may write another price vector B, for example

$$B = (6, 4, 5, 3, 5).$$

In this way, we can quickly compare the relative cost of each item at the two stores.

We may also combine the two price vectors A and B to form the rectangular array, which we call P,

$$P = \begin{pmatrix} 5 & 3 & 6 & 4 & 5 \\ 6 & 4 & 5 & 3 & 5 \end{pmatrix}.$$

Note that in P, the first row is the row vector of prices for store A and the second row, the row vector of prices at store B. Such a rectangular array of numbers is called a *matrix*. The matrix P above is said to be a 2 by 5 (written 2×5) matrix, since it has two rows and 5 columns. We may also consider each of the vectors A and B as a 1 by 5 matrix. Note that the order in which we write the numbers, called *entries*, of a matrix is important, since in our example the numbers in the columns give unit prices for the particular fruits in the order listed: oranges, lemons, pears, bananas, and peaches.

It is possible and useful to develop an algebra of matrices, and this is the subject matter of a study known as Matrix Theory. In this booklet we shall have use of only the multiplication of matrices, and therefore we shall restrict ourselves in this appendix to this operation of matrix algebra.

To introduce the operation of multiplication of matrices, we shall consider the following problem.

PROBLEM 1. Mr. Jones wants to buy 6 oranges, 3 lemons, 4 pears, 3 bananas, and 4 peaches. What is his bill at store A for this purchase? at store B?

Referring to the price vector for A, we may compute Mr. Jones's bill at store A immediately as

$$5 \cdot 6 + 3 \cdot 3 + 6 \cdot 4 + 4 \cdot 3 + 5 \cdot 4 = 95 \text{ cents} \tag{1}$$

and at store B as

$$6 \cdot 6 + 4 \cdot 3 + 5 \cdot 4 + 3 \cdot 3 + 5 \cdot 4 = 97 \text{ cents.} \tag{2}$$

We may also indicate the method of arriving at these answers by the following device. We first write the set of purchases Mr. Jones is to make as a *column vector* J,

$$J = \begin{pmatrix} 6 \\ 3 \\ 4 \\ 3 \\ 4 \end{pmatrix}.$$

We have used a column vector for the quantities of Mr. Jones's purchases to distinguish it from the price vectors A and B, which were written as row vectors. The computation used in Eq. (1) may be indicated as the product of the row vector A and the column vector J; thus

$$(5, 3, 6, 4, 5) \cdot \begin{pmatrix} 6 \\ 3 \\ 4 \\ 3 \\ 4 \end{pmatrix} = 5 \cdot 6 + 3 \cdot 3 + 6 \cdot 4 + 4 \cdot 3 + 5 \cdot 4 = 95. \qquad (3)$$

In Eq. (3), we have multiplied each element of the row vector A by the corresponding element of column vector J and have then added the results. This method of multiplying a row vector and a column vector is called *row-by-column multiplication.*

We may similarly find the answer obtained in Eq. (2) above as the product of the row vector B and the column vector J. As a matter of fact, we may indicate in one matrix equation the results for both stores A and B as

$$\begin{pmatrix} 5 & 3 & 6 & 4 & 5 \\ 6 & 4 & 5 & 3 & 5 \end{pmatrix} \begin{pmatrix} 6 \\ 3 \\ 4 \\ 3 \\ 4 \end{pmatrix} = \begin{pmatrix} 95 \\ 97 \end{pmatrix}. \qquad (4)$$

Here we have written the 2×5 matrix P, representing unit prices at stores A and B, and we define the matrix product of P and the column vector J to be the 2×1 matrix obtained by using row-by-column multiplication of the first row of P by the column vector J, then the second row of P by the column vector J.

Finally, we could use the above idea of matrix multiplication to find answers simultaneously to the questions of the cost of purchases made at stores A and B, respectively, not only by Mr. Jones, but also by another person, Mr. Smith, whose purchases are given by the column vector

$$S = \begin{pmatrix} 5 \\ 3 \\ 6 \\ 2 \\ 4 \end{pmatrix}.$$

We shall leave it to the reader to check that the matrix product

$$\begin{pmatrix} 5 & 3 & 6 & 4 & 5 \\ 6 & 4 & 5 & 3 & 5 \end{pmatrix} \begin{pmatrix} 6 & 5 \\ 3 & 3 \\ 4 & 6 \\ 3 & 2 \\ 4 & 4 \end{pmatrix} = \begin{pmatrix} 95 & 98 \\ 97 & 98 \end{pmatrix} \tag{5}$$

gives the two previous answers for Mr. Jones in the first column and those for Mr. Smith in the second column of the result. In obtaining in Eq. (5) this product of the 2×5 price matrix for stores A and B and the 5×2 matrix of purchases by Jones and Smith, we have used row-by-column multiplication, with the first row and first column to obtain the 95 in the result, first row and second column to obtain the 98 in the first row of the result; similarly, with the second row and first column to obtain the 97, and second row and second column to obtain the 98 in second row of the result.

It should be clear from the above discussion that the method of row-by-column multiplication of matrices depends on the two factor matrices having appropriate "shape" or dimensions; there must be just enough elements in each row of the first matrix to be used to multiply elementwise with the elements of the columns of the second matrix. Specifically, the first matrix must have the same number of *columns* as the number of *rows* in the second matrix. It is not necessary, however, in our example that the first matrix P have only two rows; we might list prices at a third store as the third row of P and then arrive at a 3×2 product matrix, giving Jones's and Smith's bills at each of the 3 stores. Nor is it necessary that the second matrix of our example have only two columns; we could consider the purchases for four individuals instead of just Jones and Smith.

We shall give below an important property of multiplication of matrices, for which we do not give a proof.

Associative Property. If P is an $m \times n$ matrix, Q an $n \times r$ matrix, and R an $r \times s$ matrix, then

$$(PQ)R = P(QR).$$

In particular, then, if P is a square matrix, we may write P^3 for $(P \cdot P) \cdot P$ or $P \cdot (P \cdot P)$.

The exercises below should suffice to give the reader who is unacquainted with matrix theory sufficient practice to apply these principles in the discussion of Section 8 of this book.

EXERCISES

1. For Problem 1 of this appendix, add the following information. A third store C has the following unit prices: oranges 4¢, lemons 3¢, pears 5¢, bananas 4¢, peaches 6¢. Mr. Brown purchases 6 oranges, 4 lemons, 5 pears, 3 bananas, and 3 peaches, and Mr. White buys 4 oranges, 3 lemons, 4 pears, 3 bananas, and 5 peaches. Find the total bill of each of the men Jones, Smith, Brown, White at each of the stores A, B, C, using matrix multiplication.

2. Use the following 2×2 matrices

$$A = \begin{pmatrix} 1 & 2 \\ 3 & 4 \end{pmatrix}, \quad B = \begin{pmatrix} 2 & 1 \\ 6 & 5 \end{pmatrix}, \quad C = \begin{pmatrix} 2 & 4 \\ 2 & 3 \end{pmatrix}$$

to perform the indicated matrix operations:

(a) AB (b) BA (c) $(AB)C$ (d) $A(BC)$
(e) $A^2 = A \cdot A$ (f) $A^2 \cdot A$ (g) $A \cdot A^2$ (h) A^4

3. If X is the row vector $X = (0.2, 0.8)$ and P the 2×2 matrix

$$P = \begin{pmatrix} 0.7 & 0.3 \\ 0.4 & 0.6 \end{pmatrix},$$

find XP and XP^2. Verify that $XP^2 = (XP)P$.

4. Solve the matrix equation below for the vector $Y = (y, 1 - y)$:

$$YP = Y,$$

where P is the matrix of Exercise 3 above. Perform the matrix multiplication indicated on the left-hand side of the matrix equation, and then use the property of matrices (and vectors) that two matrices (vectors) are equal if and only if their corresponding entries are all equal in pairs.

5. For the vector Y obtained in Exercise 4, find YP, YP^2, YP^3, YP^{16}.

ANSWERS TO EXERCISES

Section 2

1. (a) $\frac{1}{8}$ (b) $\frac{3}{8}$ (c) $\frac{3}{8}$ (d) $\frac{1}{8}$ (e) $\frac{7}{8}$ (f) $\frac{1}{2}$
2. (a) $\frac{1}{221}$ (b) $\frac{1}{221}$ (c) $\frac{8}{663}$
3. $\frac{2}{5}$
4. (a) $\frac{1}{5}$ (b) $\frac{3}{5}$
5. $\frac{1}{35}$
6. $\frac{1}{1960}, \frac{1419}{1960}, \frac{541}{1960}$
7. $\frac{19}{495}, \frac{4}{33}, \frac{316}{495}$
8. (a) $\frac{1}{4}$ (b) $\frac{1}{6}$ (c) $\frac{1}{2}$
9. (a) $\frac{1}{4}$ (b) $\frac{1}{4}$ (c) $\frac{1}{2}$
10. $\frac{3}{7}, \frac{4}{7}$

Section 3

1. (a) $\frac{1}{7}, \frac{1}{7}, \frac{1}{7}, \frac{2}{7}, \frac{2}{7}$ (c) $\frac{1}{7}, \frac{5}{7}, \frac{3}{7}$
2. (a) $\frac{1}{6}$ (b) $\frac{1}{6}$ (c) $\frac{11}{36}$
3. $\frac{5}{6}$
4. $\frac{1}{8}, \frac{1}{2}$
5. $\frac{1}{5}$
6. $\frac{7}{20}$

Section 4

2. (a) $\frac{1}{2}$ (b) $\frac{1}{3}$ (c) No
3. (a) $\frac{1}{2}$ (b) $\frac{1}{2}$ (c) 0 (d) Increases, increases, decreases
4. (a) $P(A \cap B) = \frac{1}{2} \neq P(A) \cdot P(B) = (\frac{3}{4})(\frac{1}{2})$
   (b) $P(A' \cap B') = \frac{3}{8} = P(A') \cdot P(B') = (\frac{1}{2})(\frac{3}{4})$, yes
6. (a) Take S as the set of all 2-subsets from the 8 cards; $N = n(S) = 28$
   (b) $\frac{3}{11}$ (c) $\frac{5}{13}$ (d) $\frac{3}{7}$
7. Yes, yes, yes, yes
8. 0.73

Section 5

1. (a) $\frac{1}{12}$ (b) $\frac{1}{2}$
2. (b) $\frac{5}{21}, \frac{3}{7}, \frac{3}{7}$
3. (b) $\frac{1}{10}, \frac{3}{10}, \frac{3}{5}$; no
4. 0.04
5. (a) $1 - (\frac{364}{365})^5$ (b) 253

6. $\frac{23}{30}, \frac{15}{23}$ 7. $\frac{2}{3}$ 8. $\frac{3}{7}$ 9. (b) $\frac{1}{3}$

10. $P\{\text{two } R \mid R\} = \frac{1}{2}$, $P\{\text{two } W \mid R\} = 0$, $P\{1\ W \text{ and } 1\ R \mid R\} = \frac{1}{2}$

11. $P\{\text{two } R \mid RR\} = \frac{2}{3}$, $P\{\text{two } W \mid RR\} = 0$, $P\{1\ W \text{ and } 1\ R \mid RR\} = \frac{1}{3}$

12. $\frac{64}{89}$

SECTION 6

1. (a) $\frac{80}{243}$ (b) $\frac{211}{243}$ (c) $\frac{4}{9}$
2. (a) 0.35 (b) 0.65 (c) 0.74
3. 0.52, 0.49
4. 13
5. $\frac{27}{128}, \frac{175}{256}$
6. (a) $X = 0, 1, 2$; $Y = 0, 1, 2$

(b) $P\{X = 0\} = P\{RR\} = \frac{2}{15}$, $P\{X = 1\} = P\{WR, RW\} = \frac{8}{15}$ $P\{X = 2\} = P\{WW\} = \frac{1}{3}$, $P\{Y = 0\} = (\frac{2}{5})^2 = \frac{4}{25}$, $P\{Y = 1\} = 2(\frac{2}{5})(\frac{3}{5}) = \frac{12}{25}$, $P\{Y = 2\} = (\frac{3}{5})^2 = \frac{9}{25}$

(c) Yes, Experiment 2

SECTION 7

1. (a) $A(d) = 1 - d(d-1)/90$
2. (a) Plan 1: $A_1(d) = (1 - d/100)(1 - d/99)$
   Plan 2: $A_2(d) = 1 - d(d-1)/9900$
3. (a) 0.172 (b) 0.224
4. Let a_1 be "adopt new treatement, a_2: "defer judgment," and a_3: "discard new treatment."

(a) $P(a_1 \mid p = \frac{1}{4}) = 0.004$, $P(a_2 \mid p = \frac{1}{4}) = 0.22$, $P(a_3 \mid p = \frac{1}{4}) = 0.776$
(b) $P(a_1 \mid p = \frac{3}{4}) = 0.776$, $P(a_2 \mid p = \frac{3}{4}) = 0.22$, $P(a_3 \mid p = \frac{3}{4}) = 0.004$

SECTION 8

1. $X_1 = (0.32, 0.68)$, $X_2 = (0.42, 0.58)$

2. $P^2 = \begin{pmatrix} 0.83 & 0.17 \\ 0.34 & 0.66 \end{pmatrix}$, $P^4 \doteq \begin{pmatrix} 0.747 & 0.253 \\ 0.507 & 0.493 \end{pmatrix}$

$P^8 \doteq \begin{pmatrix} 0.686 & 0.314 \\ 0.629 & 0.371 \end{pmatrix}$, $P^{16} \doteq \begin{pmatrix} 0.668 & 0.332 \\ 0.665 & 0.335 \end{pmatrix}$

3. $X_4 \doteq (0.627, 0.373)$, $X_8 \doteq (0.657, 0.343)$, $X_{16} \doteq (0.667, 0.333)$
4. $X_4 \doteq (0.547, 0.453)$, $X_8 \doteq (0.639, 0.361)$, $X_{16} \doteq (0.665, 0.335)$
5. (a) $P = \begin{pmatrix} 0.8 & 0.2 \\ 0.6 & 0.4 \end{pmatrix}$ (b) $(\frac{3}{4}, \frac{1}{4})$ (c) 0.7, 0.74, 0.748, $\frac{3}{4}$

6. (a) $P = \begin{pmatrix} 0.4 & 0.4 & 0.2 \\ 0.2 & 0.6 & 0.2 \\ 0.2 & 0.4 & 0.4 \end{pmatrix}$ (b) $(\frac{1}{4}, \frac{1}{2}, \frac{1}{4})$

For (c) through (e), $P^2 = \begin{pmatrix} 0.28 & 0.48 & 0.24 \\ 0.24 & 0.52 & 0.24 \\ 0.24 & 0.48 & 0.28 \end{pmatrix}$

(c) 0.28, 0.48, 0.24 (d) 0.24, 0.52, 0.24 (e) 0.24, 0.48, 0.28
(f) $\frac{1}{4}$, $\frac{1}{2}$, $\frac{1}{4}$

Appendix 1

1. (a) 20 (b) 120 (c) 120
2. (a) 12 (b) 14 (c) 196
3. (a) 26^2 (b) 26^3 (c) $26^2 + 26^3$ (d) $26^2 + 26^3 = 18,252 < 20{,}000$
4. $\{a, b, c\}$; *abc, acb, bac, bca, cab, cba*
   $\{a, b, d\}$; *abd, adb, bad, bda, dab, dba*
   $\{a, c, d\}$; *acd, adc, cad, cda, dac, dca*
   $\{b, c, d\}$; *bcd, bdc, cbd, cdb, dbc, dcb*

Each list contains the same set of arrangements.

5. (a) 720 (b) 126 (c) 126
   (d) 126 (e) 1000 (f) 1
   (g) 10! = 3,628,800 (h) 16 (i) 16
   (j) 1,999,000 (k) 2520 (l) 40
   (m) 35
6. (a) 70 (b) 35 (c) 20 (d) 15 (e) 20
7. 2520
8. (a) $\binom{10}{2} = 45$

   (b) $\binom{3}{2} \Big/ \binom{10}{2} = \frac{1}{15}$, $\binom{3}{1}\binom{7}{1} \Big/ \binom{10}{2} = \frac{7}{15}$,

   $\binom{7}{2} \Big/ \binom{10}{2} = \frac{7}{15}$
9. (a) $10 \cdot 10 = 100$
   (b) $(3)(3)/300 = \frac{9}{100}$, $(3 \cdot 7 + 7 \cdot 3)/100 = \frac{42}{100}$, $(7)(7)/100 = \frac{49}{100}$

10. $$\frac{n!}{r!(n-r)!} + \frac{n!}{(r-1)!(n-r+1)!} = \frac{n!(n-r+1+r)}{r!(n-r+1)!}$$
$$= \frac{(n+1)!}{r!(n-r+1)!}$$

Appendix 2

1. $$\begin{pmatrix} 5 & 3 & 6 & 4 & 5 \\ 6 & 4 & 5 & 3 & 5 \\ 4 & 3 & 5 & 4 & 6 \end{pmatrix} \begin{pmatrix} 6 & 5 & 6 & 4 \\ 3 & 3 & 4 & 3 \\ 4 & 6 & 5 & 4 \\ 3 & 2 & 3 & 3 \\ 4 & 4 & 3 & 5 \end{pmatrix} = \begin{pmatrix} 95 & 98 & 99 & 90 \\ 97 & 98 & 101 & 90 \\ 89 & 91 & 91 & 87 \end{pmatrix}$$

2. (a) $\begin{pmatrix} 14 & 11 \\ 30 & 23 \end{pmatrix}$ (b) $\begin{pmatrix} 5 & 8 \\ 21 & 32 \end{pmatrix}$ (c) $\begin{pmatrix} 50 & 89 \\ 106 & 189 \end{pmatrix}$ (d) $\begin{pmatrix} 50 & 89 \\ 106 & 189 \end{pmatrix}$

(e) $\begin{pmatrix} 7 & 10 \\ 15 & 22 \end{pmatrix}$ (f) $\begin{pmatrix} 37 & 54 \\ 81 & 118 \end{pmatrix}$ (g) $\begin{pmatrix} 37 & 54 \\ 81 & 118 \end{pmatrix}$ (h) $\begin{pmatrix} 199 & 290 \\ 435 & 634 \end{pmatrix}$

3. $XP = (0.46, 0.54)$, $XP^2 = (0.538, 0.462)$

4. $Y = (y, 1 - y) = (\frac{4}{7}, \frac{3}{7})$

5. $YP = Y$, $YP^2 = (YP)P = YP = Y$, $YP^3 = (YP^2)P = YP = Y$, $YP^{16} = Y$. That is, $(\frac{4}{7}, \frac{3}{7})$ is the answer in each case.

INDEX

ABCDE6987

THE AUTHOR

Grace E. Bates, Professor of Mathematics at Mount Holyoke College, received the Ph.D. degree from the University of Illinois in 1946. She has had wide teaching experience both on the college and secondary levels and has taught at Sweet Briar College, Dartmouth College, and The University of California at Berkeley. During 1963–1964, Dr. Bates served as Visiting Lecturer under the Mathematical Association of America Secondary School Lectureship Program. In 1965, she was Chairman of the Northeastern Section of the Mathematical Association of America and is presently a member of the American Mathematical Society, the Institute of Mathematical Statistics, the Canadian Mathematical Congress, and the Association for Teachers of Mathematics in New England, of which she is currently President of the Connecticut Valley Section. She is also a member of Phi Beta Kappa and Sigma Xi. Dr. Bates is coauthor with Johnson, Lendsey and Slesnick, of *Intermediate Algebra,* published by Addison-Wesley and, with Fred Kiokemeister, of *The Real Number System.*

This text, requiring only a background of high school algebra, presents an axiomatic development of probability theory as a model for experiments resulting in one of a finite number of outcomes. It can be used by the college freshman or sophomore to prepare an elementary basis for more abstract courses in probability and can serve as the primary text for a probability unit in National Science Foundation Institutes for secondary school mathematics teachers. The book contains two sections on applications, one on decision theory (acceptance sampling and testing hypotheses, binomial case) and one on the Markov Chain model. The notion of first specifying a sample space of an experiment or of building sequentially such a space is emphasized throughout the book. Two appendices are also included on counting problems and multiplication of matrices for the benefit of the student unfamiliar with these ideas.

ADDISON-WESLEY PUBLISHING COMPANY

READING, MASSACHUSETTS • PALO ALTO • LONDON • DON MILLS, ONTARIO

PRINTED IN U.S.A.

0405